Assessment in Science

A Guide to Professional Development and Classroom Practice

Edited by

Daniel P. Shepardson

Purdue University,
West Lafayette, IN, U.S.A.

KLUWER ACADEMIC PUBLISHERS

DORDRECHT / BOSTON / LONDON

A C.I.P. Catalogue record for this book is available from the Library of Congress.

ISBN 0-7923-7093-7 (HB)
ISBN 0-7923-7094-5 (PB)

Published by Kluwer Academic Publishers,
P.O. Box 17, 3300 AA Dordrecht, The Netherlands.

Sold and distributed in North, Central and South America
by Kluwer Academic Publishers,
101 Philip Drive, Norwell, MA 02061, U.S.A.

In all other countries, sold and distributed
by Kluwer Academic Publishers,
P.O. Box 322, 3300 AH Dordrecht, The Netherlands.

Printed on acid-free paper

TABLE OF CONTENTS

ACKNOWLEDGEMENTS

The writing and editing of the chapters in this book was an enormous undertaking and a complex task; however, the editorial assistance and suggestions provided by Professor Susan J. Britsch, Literacy and Language Education, Department of Curriculum and Instruction, Purdue University was invaluable in bringing this book to fruition. If it were not for the efforts of the contributors in writing chapters for this book there would be no book. I greatly appreciate and thank the contributions of these individuals to this project.

DEDICATION

I wish to dedicate this book to my parents for their support and encouragement to pursue my educational and academic interests. I also dedicate this book to Professor Gerald Krockover, Science Education, Purdue University for his mentorship in science education. I wish to acknowledge the many excellent teachers I have had the opportunity to collaborate with over the years, for without their participation in my professional development projects I would not have been able to develop my understandings about professional development, teaching, learning, and assessment.

DANIEL P. SHEPARDSON

1. INTRODUCTION TO ASSESSMENT IN SCIENCE: A GUIDE TO PROFESSIONAL DEVELOPMENT AND CLASSROOM PRACTICE

The purpose of this book is to present ideas, raise issues, share tools and techniques, and stimulate thinking about assessment in science classrooms. With the aim of enhancing the professional development of teachers in science assessment, the book includes examples of professional development strategies and classroom assessments. The book is pragmatic in that it presents practical tasks, professional development models and activities, and tools and templates for planning and conducting assessment in science   Although the book does not describe or debate assessment policy or large-scale assessments, it does build on the national reform documents in science education: the *National Science Education Standards* (National Research Council [NRC], 1996) and the *Benchmarks for Science Literacy* (American Association for the Advancement of Science [AAAS], 1993).

The audience for this book includes staff developers, classroom teachers, and school administrators. For staff developers and school administrators the aim of the book is to provide ideas for conducting successful professional development programs in science assessment that facilitate teacher change. The book alerts such individuals to the constraints and issues surrounding the professional development and implementation of assessment in science classrooms. For classroom teachers and school administrators the aim is to provide examples of classroom practice, models and templates of classroom assessments, ideas for developing assessment tasks, and insight into issues surrounding classroom assessment in science. The goal is to initiate thinking about assessment as an integral system within the science classroom versus an afterthought or an addition to instruction.

Professional development in science assessment that emphasizes an understanding of assessment and the analysis of student work fosters teachers to change instruction as the student work provides evidence for the need to change. Thus, changing science assessment practice has the potential to change what is taught and learned in science classrooms. This in turn changes what students learn, what students know and can do in science. Professional development programs that

1

Daniel P. Shepardson (ed.), Assessment in Science, 1—5.
© 2001 *Kluwer Academic Publishers. Printed in the Netherlands.*

build on teachers' understandings of assessment, the context of the science classroom, incorporate teacher-staff collaboration and teacher reflection, and provide ongoing support are more likely to be successful in changing teachers' assessment practice.

To this end, the book is divided into two sections. The first addresses ideas and issues surrounding the professional development of teachers in science assessment. The chapters in this section focus on the use of the *National Science Education Standards* (NRC, 1996) and the *Benchmarks for Science Literacy* (AAAS, 1993) in the professional development of teachers. It also presents practical activities for promoting teacher reflection on practice. The chapters in this section explore the constraints teachers may face when changing their assessment practice. The second section contains case studies of classroom assessment practice, presenting ideas for the development of classroom assessment tasks and models of classroom assessment practice. In section two classroom teachers also articulate their views of assessment in science and tell their stories about the development and use of assessment in their classrooms.

Chapter 2 then introduces professional development in science assessment by describing a professional development framework and several activities that operationalize the NRC (1996) standards in professional development. A professional development framework that builds on Shulman's (1987) model of pedagogical reasoning and action is also presented. Examples of professional development activities illustrate the use of Shulman's model to engage teachers in thinking about assessment in science. The role of action research in facilitating teachers' reflection on their classroom assessment practice is also presented. The chapter closes by discussing tools for promoting teachers' thinking and development of assessment, including the use of the *Benchmarks for Science Literacy* (AAAS, 1993) and *Science for all Americans* (AAAS, 1990). Finally, Chapter 2 addresses the technology of assessment and teachers as consumers of assessment.

In Chapter 3 Gummer and Shepardson discuss professional development strategies for introducing assessment to teachers of science, building a consensual understanding of assessment among teachers, and using the *National Science Education Standards* (NRC, 1996) as a tool in professional development. The chapter describes the assessment evaluation matrix derived from the NRC (1996) Standards by teachers as apart of the Teacher Enhancement Through Alternative Assessment Task Development project (Shepardson, 1996). In Chapter 4 Gummer and Shepardson provide an overview of the change process and the difficulties teachers may encounter in changing their classroom assessment practice. This chapter identifies specific issues and constraints that should be considered by staff developers as they plan professional development programs.

The chapter titled, "Thinking about Assessment: An Example from an Elementary Classroom," presents a case study of how one elementary teacher changed her assessment practice as a result of reflection on her teaching practice and her interactions with a colleague. The chapter describes issues associated with changing classroom assessment practice, ways of thinking about classroom assessment, along with examples of classroom assessments. The role of science education research in assessment task development is also explored.

Chapter 6 articulates an assessment framework for planning science assessment at the classroom, school, or district level. Shepardson and Gummer propose the science domains and the performance categories that should be considered in the development of science assessments. They link national reform documents (e.g. *National Science Education Standards*, NRC, 1996; *Benchmarks for Science Literacy*, AAAS, 1993) to their proposed assessment framework and build from national and international assessment frameworks (e.g., NAEP, TIMSS).

The second section of the book, Chapters 7 through 14, specifically focus on science assessment in classrooms, providing examples of classroom practice and assessment tasks. In Chapter 7 Britsch describes the assessment of science learning through the talking, drawing, and writing of very young children. Britsch illustrates this assessment through children's work examples from a kindergarten classroom experience in which children dissolved substances. Britsch also discusses the importance of teacher self-assessment for improving instruction for determining the appropriateness of science experiences for young children.

In Chapter 8 Shepardson and Britsch provide an overview of assessment, evaluation and grading, planning for classroom assessment, and assessment as profiling. The heart of the chapter is its presentation of ideas and techniques for assessing children's graphic products and children's planning and conducting of investigations. The chapter closes with an examination of children's self-produced journals as a tool for teaching, learning, and assessing.

Main's chapter, "Assessing Children's Science Learning and Process Skills in the Elementary Classroom," (Chapter 9) characterizes how the *Benchmarks for Science Literacy* (AAAS, 1993) and the *National Science Education Standards* (NRC, 1996) served as guides for her teaching and assessment practice. She describes a sink and float assessment task--a practical task--and the scoring rubric used. Main also describes informal assessment, child interviews as assessment, practical task assessment stations, and observational checklists. For each she provides suggestions for successful development and implementation. She highlights the importance of revising assessment tasks and scoring guides based on student responses and the need to implement change in a manageable fashion.

In Chapter 10, Edwards describes several pedagogical-assessment activities that she uses with her seventh-grade students in her life science classes. Primarily discussing the assessment task titled, "What is the Relationship between Abiotic and Biotic Factors," the scoring guide used to assess student performance on the task, and student work samples. Edwards describes how she used a diagnostic assessment to learn more about her students' prior ideas and the influence of science education research literature on her assessment practice. She also overviews a pedagogical-assessment activity that serves as a summative assessment task. Edwards concludes the chapter by talking about standards-based assessment and the importance of alternative assessment in science education reform.

Jackson shares examples of assessment tasks from a comprehensive science assessment for the seventh-grade in her chapter, "Multidimensional Assessment of Student Performance in Middle School" (Chapter 11). Jackson articulates the importance of developing assessment tasks that align with instruction and curriculum, using the *Benchmarks for Science Literacy* (1993) in identifying the standards to be assessed. Jackson details three different assessment tasks and their scoring rubrics, illustrates with student work examples, and reflects on the students' work. She emphasizes the importance of revising assessment tasks and scoring rubrics based on student responses as well as interviewing students to determine their perspective about the assessment tasks.

Chapter 12, "Developing and Using Diagnostic and Summative Assessments to Determine Students Conceptual Understandings in a Junior High School Earth Science Classroom," a teacher's change in thinking about scientific literacy, achievement and learning, and the use of diagnostic and summative assessments in science. Leuenberger shares the diagnostic and summative assessment tasks he developed to assess students' conceptual understanding about convection currents in air. He first shares student work examples and the scoring rubrics used to assess student work and then shares his insights into the development and use of a summative assessment, again providing examples of student work and scoring rubrics. Leuenberger concludes the chapter by restating the importance of developing diagnostic assessments that address major conceptual goals that build on previously taught concepts.

In "Alternatives to Teaching and Assessing in A High School Chemistry Classroom: Computer Animations and Other Forms of Visualization" (Chapter 13) Flick illustrates the importance of aligning assessment and instruction, and utilizing alternative approaches to teaching and assessing in a high school chemistry course. The chapter emphasizes Flick's use of computer animation and visualization techniques as assessment tools. Her instruction and assessment are guided by the *Benchmarks for Science Literacy* (AAAS, 1993) to articulate what high school seniors should know about the structure of matter. Flick first details the goals of her assessment techniques and then describes two instructional units. The first deals with chemical reactions, classifications, and representation; the second covers chemical bonding, valence electrons, and molecular shapes. Flick describes her revision of the units based on assessment data, using classroom assessment to both inform pedagogy and evaluate students.

Emery's chapter tackles perhaps one of the most controversial issues of alternative assessment: authentic assessment. In Chapter 14, Emery first describes what authentic science assessment might look like based on the reform documents in science education (e.g., *Benchmarks for Science Literacy*, AAAS, 1993; *National Science Education Standards,* NRC, 1996) and then introduces his perspective as a classroom science teacher. Emery then walks the reader through a synopsis of his personal growth and development in using authentic assessment and developing scoring guides, laying out his errors and evolution. Based on this personal growth he presents suggestions for developing successful authentic assessment tasks and avoiding the pit-falls he encountered. Emery describes how his authentic assessment tasks evolved to incorporate student self- and peer-assessment, linking this to the NRC (1996) Standards and building on the notion of developing self-directed learners. Emery then shares his evaluation of and reflections about two assessment tasks. The chapter closes by noting the importance of authentic assessment in science as a means to improve students' problem-solving and higher-order thinking abilities.

The last section and chapter in the book summarizes current science assessment issues related to professional development and classroom practice. The chapter closes with a set of future needs and recommendations for improving science assessment practice and supporting science teachers in changing their classroom assessment practice.

REFERENCES

American Association for the Advancement of Science. (1993). *Benchmarks for Science Literacy.* New York: Oxford University Press.

National Research Council. (1996). *National Science Education Standards.* Washington, DC: National Academy Press.

Shepardson, D.P. (1996). Teacher enhancement through alternative assessment task development. A proposal funded by the State of Indiana Commission for Higher Education.

Shulman, L.S. (1987). Knowledge and teaching: Foundations of the new reform. *Harvard Educational Review, 57,* 1-22.

DANIEL P. SHEPARDSON

INTRODUCTION TO SECTION I: STRATEGIES AND TECHNIQUES FOR PROFESSIONAL DEVELOPMENT

This section (Chapters 2 through 6) details successful professional development strategies and activities for challenging and changing teachers' assessment practice. Issues concerning the professional development of teachers are described. The purposes of this section are to provide staff developers and teachers with practical strategies for reflecting on and changing classroom assessment practice and to inform staff developers, administrators, and teachers about classroom-based issues concerning assessment practice.

Chapter 2 outlines a professional development framework that utilizes teacher reflection as an integral component to challenging and changing teachers' assessment practice. The use of *Science for All Americans* (AAAS, 1990) and *Benchmarks for Science Literacy* (AAAS, 1993) by teachers as tools for changing practice is described in the chapter. Chapter 3 continues the professional development discussion by describing how teachers participating in a professional development program used the *National Science Education Standards* (NRC, 1996) as a tool for constructing an assessment evaluation matrix for evaluating classroom assessment practice. Additional professional development strategies and activities are described. Chapter 4 articulates the professional development issues that must be considered in the design and implementation of professional development programs for teachers. The chapter outlines the barriers to changing teacher assessment practice that must be broached in professional development programs.

Chapter 5 presents the collaborative activities and dialogue surrounding the process of changing an elementary teacher's classroom assessment practice. The case study approach provides a classroom perspective on teacher assessment practice, describing the strategies and techniques used to reflect on and change classroom assessment practice. Examples of classroom-based assessments are presented.

The last chapter (Chapter 6) of this section outlines an assessment framework for thinking about and designing assessments for use in the classroom or at the school district level. The chapter presents the science domains that should be considered when designing assessment systems and classroom assessments. A professional development scenario is presented illustrating how teachers might use the assessment framework as a tool for developing assessments that align with instruction and the national standards.

Daniel P. Shepardson (ed.), Assessment in Science, 7.

DANIEL P. SHEPARDSON

2. A PROFESSIONAL DEVELOPMENT FRAMEWORK FOR COLLABORATING WITH TEACHERS TO CHANGE CLASSROOM ASSESSMENT PRACTICE

The *National Science Education Sta*ndards (National Research Council [NRC], 1996) for professional development call for dramatic change in the nature and delivery of professional development activities. The NRC (1996) standards call for professional development to move from activities that emphasize the learning of technical skills to activities that engage science teachers in building knowledge and understanding about practice. Such activities need to be appropriately connected to the context of teachers' work in classrooms and schools. These assumptions emphasize that teachers be provided with opportunities to reflect upon their practice, to engage in dialogue with others about practice, and to collaborate with others in planning and developing curriculum, instruction, and assessment tools. It is essential that professional development activities concurrently address teachers' understandings about curriculum, instruction, and assessment. Changing instructional practice without addressing curricular issues and assessment practice, or changing assessment practice without addressing curricular issues and instructional practices, will not lead to significant change in classroom practice.

This chapter describes a professional development framework and instructional activities that model the delivery of a professional development program that aligns with the NRC standards. In addition, strategies for using *Science for All Americans* (AAAS, 1990) and *Benchmarks for Scientific Literacy* (AAAS, 1993) as tools for change, as well as other strategies and tools for reflecting on and changing assessment practice are presented. This professional development framework and strategies have been used successfully in numerous professional development programs. The framework revolves around the establishment of an intellectual community of learners that challenges teachers' understandings about practice, promotes teacher reflection upon practice, and encourages dialogue and collaboration among teachers and project staff.

The premise of the professional development framework is that learning is a sociocultural constructive process. Thus, teachers must construct their own purposes and meanings for instruction and assessment as it relates to their classroom communities and school cultures. Further, participants bring with them what Shulman (1987) referred to as "wisdom-of-practice" knowledge: existing knowledge about science instruction, curriculum, and assessment constructed from experiences

9

Daniel P. Shepardson (ed.), Assessment in Science, 9—37.
© 2001 *Kluwer Academic Publishers. Printed in the Netherlands.*

in classrooms. These existing conceptions contribute significantly to the shared meanings constructed by both participants and project staff.

A major difficulty facing curricular and instructional reform is the mismatch between the curricular and instructional reform efforts and existing classroom assessment practice. In order to successfully change practice, professional development programs must concurrently address the curricular, instructional, and assessment issues associated with reform. Few educational reform efforts address these three dimensions--curriculum that provides the content, instruction that mediates the content, and assessment that evaluates the process and which is grounded in the context of being educational (Eisner, 1985). These three dimensions form a knowledge base for understanding teaching, assessing, and learning, or pedagogical content assessment knowledge. The assumptions underlying pedagogical content assessment knowledge are:

- The science content we know influences the ways we teach and assess, as well as what we teach and assess.
- The pedagogy we know influences the science content we teach, how we teach, and what we assess.
- What we know about assessment influences what we assess, how we assess, how we teach, and the science content we teach.

On this view, assessment functions as both a way of teaching and as a way of learning.

BACKGROUND ABOUT PROFESSIONAL DEVELOPMENT

Teachers tend to base their curricular and instructional decisions on their individual backgrounds, interests, and experiences (Zahorik, 1984) and on the culture of the school (Sarason, 1971). As a result, there often exists a discrepancy between the innovation and the teachers' perception of the innovation and its implementation in the reality of the classroom (Olson, 1980, 1981; Doyle & Ponder, 1977). Addressing this innovation-classroom gap is a critical factor in changing classroom practice.

To promote change in practice, teachers must not only be aware of their beliefs and of current practice, but must also have alternatives to practice available (Nespor, 1987); however, providing alternatives to practice alone will not significantly change practice. The professional development experience must challenge teachers' assumptions and understandings about the teaching-assessing-learning process, and provide alternatives to practice. By subjecting current practice to critical reflection, change transforms the way practice is experienced and understood (Robottom, 1987). This critical reflection creates cognitive dissonance so that new practice is connected to the context of the teachers' classroom (Huberman, 1995). New practice evolves from teachers' critical reflection on the innovation, not from the authority-laden wisdom of experts (Olson, 1981). In this way, the professional development process stimulates new ideas about practice and the negotiation of a consensual understanding of practice among peers and project staff (Taylor, 1993).

Simmons (1984) stressed that teacher change occurs through the active participation of teachers as professionals. Teachers become active participants by reflecting upon an innovation in terms of their school and classroom cultures and by engaging in collaborative dialogue about the innovation and about classroom practice (Richardson, 1990). This active participation supports teacher change as teachers reconstruct their understandings and beliefs about practice and about the innovation both privately and publicly (Fenstermacher, 1979; 1986). The innovation then becomes relevant to their existing curricular, instructional (Feely, 1986), and assessment beliefs.

THE PROFESSIONAL DEVELOPMENT MODEL

A professional development model that promotes reflection upon practice, collaborative dialogue, and the negotiation of a consensual understanding of practice follows Shulman's (1987) model of pedagogical reasoning and action. This general framework promotes an intellectual basis for teaching and involves:

- Modeling ways of understanding teaching and assessing by project staff.
- Comprehending ways of understanding teaching and assessing through collaborative dialogue among staff and participants.
- Transforming and applying ways of understanding teaching and assessing in the classroom context of the participant.
- Reflecting upon the classroom application of ways of understanding teaching and assessing through writing.
- Sharing ways of understanding teaching and assessing with peers and project staff through small-group, collaborative dialogue.
- Reflecting upon collaborative dialogue, leading to a "new" comprehension of teaching and assessing.
- Developing instruction and assessment tools based upon the new comprehension.

This process establishes what McNiff (1993) calls a dialogical community where dialogue is ongoing and where participants and staff are equal peer-practitioners. This means that each is concerned about the other, about each other's understanding of practice, and about each other's progress. The teacher comments below illustrate the importance of establishing a community of learners, that supports reflective dialogue:

> [The professional development activities] has successfully prompted, encouraged, and demanded at times much more discussion between my colleagues and me. The discussions have been very thought provoking and we have struggled, agreed, and disagreed (sometimes vehemently) over implementation of some new ideas [Bill].

The collaboration with other teachers in my corporation [i.e., school districts] was most valuable in that we were able to act as sounding boards with each other. We could talk about what we can do, what we should do, and what we would like to do. . . . The collaboration between teachers has given us all new ideas, shown us what we're doing right, and what we're doing wrong [Ray].

Both Bill's and Ray's responses demonstrate the importance of collaborative group dialogue in promoting teacher discussion about practice, providing insight into "what we can do, what we should do, and what we would like to do," in Ray's words. Through continued dialogue, participants come to know and understand ways of teaching and assessing that leads to change in practice.

Hubermann (1991) concluded that successful teacher change encourages and supports teachers in their tinkering with classroom practice. Teachers develop new ways of teaching and assessing and then reflect upon those practices in the context of their classroom communities and school cultures. Reflection activities involve teachers in praxis--doing, reflecting, learning, and changing. To promote praxis, reflection activities incorporate the elements Stallings (cited in Fullan, 1990) identified as being successful for facilitating change in practice:

- Teachers analyze their current practice, becoming aware of the need for change.
- Teachers modify ideas and innovations to fit their classroom communities and school cultures.
- Teachers evaluate the innovation and the impact on both their practice and their students.
- Teachers observe and reflect upon other teachers' implementation and understanding of the innovation.
- Teachers reflect upon and engage in dialogue about their successes and failures with the innovation.
- Teachers critique and reflect upon their own practice through the use of videotape.

EXAMPLE OF PROFESSIONAL DEVELOPMENT ACTIVITIES

This professional development framework has been used to develop teachers' understandings about assessment and their assessment practice. The four examples presented below are representative of the process, and are intended to provide examples for collaborating with teachers. They illustrate the instructional activities and reflection assignments involved in the professional development model. In general, the modeling and comprehending phases occur in whole group workshop settings; the transforming, applying, and reflecting phases take place in participants' classroom environments; and the sharing, reflecting on collaborative dialogue, and developing phases reunite teachers in a whole group workshop setting. Each professional development block takes place in a three-day sequence, building on the previous sequence.

Constructing a Teaching and Assessment Profile for Reflecting on Practice

The design of the initial reflection activity assisted participants in constructing a profile of their teaching and assessment practices (see Appendix A). This reflection assignment was assigned prior to the workshops on teaching and assessment. The purpose of the activity was twofold: first, to engage participants in coming to know and understand their current practices and second, to establish a knowledge base for changing practice. An additional purpose was to solidify a cadre of teachers committed to understanding and changing their practice. For without teacher commitment, change in practice is unlikely, yet many educational reform processes have neglected this factor (Eisner, 1994).

Although teacher-participants found the teaching and assessment profile difficult to construct and time consuming, they also viewed it as interesting and rewarding, contributing to their understanding of current practice. Bill's and Peter's views are representative:

> From the very start I have had some very valuable teaching lessons. When we videotaped our instructional unit (our first assignment) I learned that as progressive as I thought I was, I teach in a traditional way [Bill].

> In order to make any changes, one has to know where they are and what they are presently doing. The process was an eye opener for me because it made me step back and think about what, how, and why I was teaching the material and to think about what could be done differently [Peter].

Bill and Peter illustrate the importance of reflection to establish understanding of current practice and to provide a knowledge base for changing practice. As Peter says, "In order to make any changes, one has to know where they are and what they are presently doing."

These profiles of practice contained samples of teachers' classroom instruction and assessments, videotapes of practice, and student work examples. Staff also engaged teachers in thinking about assessment practice by presenting the Matrix of Classroom Assessment Practice (Figure 1). The matrix is composed of opposing assessment constructs or dimensions that reflect a continuum in classroom assessment practice. It offers a diverse range of assessment constructs and questions for thinking about assessment practice; for example, is the teacher's classroom assessment on demand, ongoing, or somewhere in between? The matrix also includes a section on assessment formats (e.g., quizzes, laboratory practicals). As a comprehension activity, teachers were provided a sample assessment profile to analyze using the matrix and to discuss in their small groups. Based on these two activities, teachers modified the matrix of assessment practice to reflect their own understandings and perspectives. Questions that teachers generated about assessment practice were also used to guide their analysis and reflection, for example:

- What is the purpose of these assessments?
- What do these assessments measure about student ability or performance?

- What other assessments might be used?
- How are the assessments administered to students?
- What are the advantages and disadvantages of these assessments?
- How might you change or improve these assessments?
- How are these assessments linked to instruction?

	Always	Mostly	Balanced	Mostly	Always	
On Demand						Ongoing
Individual						Group
Closed-ended						Open-ended
Rote						Authentic
Product Driven						Process and Product
Single Discipline						Interdisciplinary
Single Trait						Multitrait
Unassisted						Assisted
Written						Oral
End of Unit						Throughout Unit
Summative						Diagnostic
Answer Only						Reasoning Processes
Fact-based						Performance-based

Assessment Formats Used (e.g., Quizzes, Tests, Lab Reports, Oral Presentations):

Questions about Classroom Assessment Practice:

Figure 1. Matrix of Classroom Assessment Practice

As a reflection assignment teachers were next instructed to transform and apply the matrix to the analysis of their assessment practice (profile). Both the matrix and the teacher generated questions facilitated teachers' reflections. They next used the matrix to tally their classroom assessments; tallies falling primarily on the left side of the matrix indicate a more traditional approach to student assessment. If the frequency of tallies falls toward the middle (balanced) or right side of the matrix, the teacher's assessment practice reflects a more alternative approach to assessment. Returning to the workshop setting, teachers share their reflections on their current assessment practice, engage in dialogue and discuss ways of improving or changing their assessment practice.

Introduction to Alternative Assessment

To promote an intellectual and reflective atmosphere, assessment strategies were modeled to provide a basis from which understandings about assessment could be constructed and socially negotiated. The following assessment techniques were modeled during an instructional sequence on pendulums: (a) written responses to open-ended questions used prior to instruction; (b) observational checklists used during the instructional activities; (c) interview questions also used during the instructional activity; and (d) a traditional, multiple-choice test administered at the end of the pendulum activity. This pedagogical-assessment experience served as the foundation for discussing and analyzing assessment, with participants and staff expressing personal beliefs and understandings about assessment. In small groups participants also discussed, analyzed, and reflected upon the assessments from the perspectives of both the teacher and student. As part of the modeling and comprehending process, teachers were introduced to the technology of assessment (see later section of this chapter) to be used in the transforming, applying, and reflecting assignment.

Participants transformed, applied, and reflected upon assessment by reading scholarly articles on assessment, developing and implementing assessment prototypes, reflecting upon their assessment practice, and videotaping practice and collecting student assessment artifacts. The transforming, applying, and reflecting assignment followed the workshop and required participants to develop two alternative assessment prototypes (see Appendix B). The purpose of this activity was to assist participants in constructing an understanding of and a need for alternative assessment techniques. Specifically, the assignment was to: (a) provide participants with practice in developing and implementing alternative assessment techniques, (b) assist participants in understanding that different assessment techniques provide different pictures of student performance, (c) demonstrate the need for alternative assessment techniques, and (d) promote collaborative dialogue about assessment practice.

At the next workshop, each teacher shared a videotape of practice, collected student artifacts (work examples), the developed assessment tasks, and the self-reflections. These served as focal points for collaborative dialogue that enabled participants to share their assessment tasks and practice in small groups of staff and peers. The small-group sharing led to further reflection upon assessment practice and to a new comprehension of assessment for both presenter and group members. Developing/revising assessment tasks occurred in small groups of peers and resulted in the application of new ways of understanding assessment. The quotations below reflect the responses of participants on the alternative assessment assignment:

> The process of trying to come up with the different forms of assessment was worthwhile in that it caused me to stop and really think about how to do something different from what I had been doing [Jane].

> Having a traditional versus Alternative Assessment assignment was beneficial in that it showed very well that some students did well and some students didn't do well on the tests because of the format. The Alternative Assessment showed that students think very differently from each other, and that all students who received the same information at the same time all processed the information differently [Bill].

Jane's response illustrates the importance of the alternative assessment activity in promoting teachers' thinking about assessing their students differently. Bill's reply emphasizes not only the need to think about assessing students differently but also, and perhaps more importantly, the different picture it is possible to construct of students by using different assessment approaches. When teachers come to know what their students can and cannot do it prompts teachers to reconsider their existing ideas and understandings about practice (Darling-Hammond, Ancess, & Falk, 1995).

The three teacher examples (Mary, Nick and Sue, and Yolanda) that follow illustrate the importance of engaging teachers in the development and implementation of alternative forms of assessment. The teachers' comments about their initial assessment tasks reveal the importance of engaging in reflection about new assessment tools. Reflection encourages teachers to identify the benefits and difficulties of their assessments and to identify ways of extending or improving their assessment techniques. The reflection component appears to promote a more positive view of the change process; teachers may even evaluate difficulties with the assessment more positively.

Mary's ObservationalChecklist

Mary developed an observational checklist to assess her students during an owl pellet investigation. Instead of general science process skills, the observational checklist contained performance categories specific to the investigation: separates bones into categories, able to use tweezers and blunt nosed needle, and cleans bones thoroughly. Mary's scoring system consisted of four ratings: +, works independently; ✓, needed some assistance; -, can complete only with help; and 0, cannot complete. Mary had the following thoughts about her first alternative assessment tool: "I see this format having possibilities for many activities . . . I could decide for each activity what I would expect to observe as the students conduct their investigation." This statement indicates that Mary envisions ways of extending her observational checklist to assess children's performance in other science investigations; however, she also notes some difficulties with this form of assessment:

> The difficulties of alternative assessment relate to the time it takes to develop the tool. Also with the observation checklist, I found that it was not always easy to put down an accurate assessment of each student. It is also necessary to keep focused on the checklist. As I work with the students I tend to get involved or stop to assist with problems and found I don't always get to everyone.

Although Mary made note of the difficulties in using her alternative assessment, she also identified the positive aspects of alternative forms of assessment:

> [H]aving the observable skills checklist helped me to focus on those specific skills and having everything on one page made it easy to locate individual students. I prefer the observation checklist because I am looking at what a student is doing. Even though I may not always be correct, the more I use that kind of assessment, the more knowledge I have of what the student can do.

By reflecting on her assessment tool Mary identified the difficulties, providing a basis from which to revise her assessment practice. By acknowledging the positive attributes of her assessment she recognizes the value and need for alternative ways of assessing students.

Nick and Sue's Practical Task Assessment

Nick and Sue, who taught in different school districts, collaboratively developed a practical task to serve as a summative assessment for their unit on energy flow in a community. Their assessment task required students to conduct an eco-column investigation, making observations over five days and apply their understandings about energy flow to a new eco-column environment. The students also created a food chain using the organisms in their eco-column. The assessment utilized an analytic scoring rubric to assess students' content understandings, as well as their performance abilities, a multitrait assessment task. Both Nick and Sue were pleased with the assessment task, indicating:

> Students were very motivated in conducting the task, and did not think of it as an assessment; the only difficulty encountered was a timing issue with the availability of living organisms. The only planned revision is to expand or more clearly specify the science process skills to be assessed.

Yolanda's Poster Assessment

Yolanda developed a summative assessment task that required pairs of students to develop a poster illustrating the water cycle. Students were allowed to utilize any resources in the room to assist them in creating their poster; however, they had to explain ideas using their own words. The prompt for the task included a list of words that students were required to integrate into the drawing, labeling, and explaining of their posters. Yolanda's first attempt at incorporating alternative assessment into her classroom was quite successful in that, "Students seemed to do very well on the assessment." Although pleased with her students' performance, Yolanda was able to identify a gap in students' understanding, "[T]hey did not make the connection that water would be transferred from plants to animals or from animals to animals through consumption." It is unlikely that a traditional multiple-choice test would have revealed this conceptual void in students' understanding. Although Yolanda was pleased with the assessment task, she plans to change her scoring system from an analytic rubric to a holistic rubric:

> I found that the rubric was emphasizing terms to much and not the connections or relation between the words, labels, and the drawings. I basically left this out. I also think that a holistic rubric will be easier to use as I want to judge the quality of the whole product not just the words. So I will develop a holistic rubric for the next time.

For Yolanda, the assignment enabled her to construct a more detailed picture of her students' conceptual understanding. The reflection process provided her with insight into ways of revising and improving her assessment task, specifically the scoring system.

Assessment in the Context of Instruction

Professional development programs must integrate assessment and instruction in order to support teachers in changing their classroom practice (American Association for the Advancement of Science [AAAS], 1998). The states of matter example shared here is intended to illustrate a model for the planning and development of instruction and assessment. The model is used to facilitate discussion and reflection on teaching, assessment and the importance of linking assessment to instruction. The model also demonstrates the connection of children's ideas to instruction and of standards to assessment development. The model presents assessment for the purpose of diagnosing student understanding and for the purpose of evaluating student learning through formative and summative assessment. The summative assessment also illustrates an authentic assessment task. The instruction and the assessments were developed using the benchmarks related to the structure of matter for grades 6-8 as a resource:

> Atoms and molecules are perpetually in motion. Increased temperature means greater average energy of motion, so most substances expand when heated. In solids, the atoms are closely locked in position and can only vibrate. In liquids, the atoms or molecules have higher energy of motion, are more loosely connected, and can slide past one another; some molecules may get enough energy to escape into a gas. In gases, the atoms or molecules have still more energy of motion and are free of one another except during occasional collisions (The Physical Setting, Structure of Matter, p. 78).

> Heat can be transferred through materials by the collisions of atoms or across space by radiation. If the material is fluid, currents will be set up in it that aid the transfer of heat (The Physical Setting, Energy Transformation, p. 85).

A list of propositional knowledge statements and concepts were generated from these benchmarks and a concept map was developed. Based on the concept map, the following instructional activities were planned and sequenced: classifying materials, properties of liquids, properties of solids, discrepant properties in solids, properties of gases, melting solids, evaporating liquids, condensing gases, freezing liquids, and sublimation.   The science education research literature provided insight into children's understandings, leading to the identification of seven potential conceptual difficulties: 1) solids like sugar and salt may be poured, 2) solids are heavy, 3) solids do not bend or break, 4) viscous liquids do not pour well and are solid-like, 5) Jell-o is a solid, 6) Thixotropic liquids (cornstarch), and 7) phase changes.   Assessment activities were then planned to align with the instructional sequence.   This pedagogical-assessment development process modeled:

- Stating the instructional goals and objectives.
- Developing the conceptual understanding of the topic.
- Identifying potential student conceptual difficulties.
- Planning and sequencing the instructional activities based on the concept map and student understandings.
- Identifying and articulating the criteria or standards of student performance based on the Benchmarks.

Planning and developing assessments that aligned with instruction, incorporated performance standards, and that were administered at the initiation of the unit (diagnostic assessment), within each activity (formative assessment), and at the end of the unit (summative assessment).

To engage teachers in thinking about assessment in the context of instruction, teachers were presented with the concept map, instructional sequence, and potential conceptual difficulties as a model. Teachers then developed assessments that could be used within the instructional activities, both aligning and embedding assessment in the context of instruction.   Teachers developed a variety of assessments and scoring systems for the instructional activities. To engage participants in thinking about summative assessment, teachers completed two activities: first, they developed a holistic scoring rubric for an existing assessment task; next, they developed an assessment task for the structure of matter unit. This also provided teachers with practice in developing scoring systems.   To assist teachers in developing a holistic rubric, examples were shared and rewritten as analytic rubrics and analytic rubrics were rewritten to reflect holistic rubrics.

To illustrate an authentic assessment task, teachers were provided with the fudge task (see Appendix D) without the scoring rubric. The rubric was provided later. Teachers were then asked to analyze the task based on the following questions:

- What makes the task authentic?
- What concepts or processes are being assessed?
- How well would the assessment align with instruction?
- What difficulties do you foresee in implementing this assessment?
- How challenging would this assessment be for students?
- Is this an example of a good assessment task?

Teachers were then provided with the scoring rubric and were asked:

- Does the scoring rubric assess what the task asks students to do?
- Is the scoring rubric aligned with state and national standards?
- Does the scoring rubric allow inferences to be drawn about student performance?
- Does the task assess conceptual understanding, reasoning skills, science process skills, and/or procedural knowledge?

At the conclusion of this activity and the workshop experience, teachers were required to develop an instructional unit that embedded assessment into their teaching. This assignment reflects the transforming, applying, and reflecting components of the framework. Teachers were required to develop and implement the assessment tasks, videotape practice, collect student assessment artifacts, and reflect upon their assessment practice. Teachers then shared their examples of teaching and assessing and their reflections in small groups.

ISSUES ABOUT SCORING SYSTEMS

A difficulty encountered by teachers in the construction of assessment tasks is the development and use of a scoring system. To address this difficulty, additional time and experience is needed. The following professional development sequence has been used to assist teachers in better understanding scoring systems, particularly scoring rubrics. To model the development of a scoring rubric the "Beaker of Ice Task" was administered to the teachers. They brainstormed a list of concepts that could be assessed based on the task and then consulted the benchmarks (AAAS, 1993) to identify the performance standards. Next, teachers developed an analytic scoring rubric for the task based on the list of concepts and the benchmarks. This provided teachers with practice in developing an analytic scoring rubric. The different scoring rubrics were then presented and discussed by participants.

An analytic scoring rubric was then constructed based on whole-group consensus. Participants used this to score each group member's response, ensuring that each response was scored by at least three different raters. Group members then compared their ratings of other teachers' performance. Issues of instructional context and rater and interrater reliability were addressed in light of the scored responses. This process was repeated when the teachers reconstructed the analytic rubric as a holistic rubric. The process itself reflected the modeling and comprehending phases of the professional development framework. The transforming, applying, and reflecting phases were conducted in the context of the teachers' classrooms when they were required to develop an assessment task that contained both an analytic and a holistic scoring rubric. Teachers then administered the assessment task to at least one class and collected data similar to those in the "Introduction to Alternative Assessment" reflection assignment. Teachers then shared their reflections, assessment task and scoring system, and student work examples with a small group of peer participants and engaged in collaborative dialogue about the assessment tasks and scoring rubrics.

ACTION RESEARCH ASSIGNMENT: INVESTIGATING CLASSROOM ASSESSMENT PRACTICE

The purpose of the action research assignment was to engage participants in evaluating the impact of the professional development program on their classroom practice. Participants designed and implemented an action research project to answer a question of interest to them. One aspect of the assignment included investigating assessment practice. The action research assignment provided participants an opportunity to: (a) link theory to classroom practice, (b) become doers of research versus consumers of research, (c) work collaboratively toward shared goals, and (d) contribute to the knowledge base for understanding assessment practice (Perry-Sheldon & Allain, 1987).

Presented here are the findings and reflections of two participants, Bill and Laura, who collaborated on investigating, in their words, "The success of students on traditional assessment questions compared to their success on alternative assessment questions." Bill and Laura designed a summative assessment that contained traditional and alternative question formats. For these two teachers, traditional assessment questions included fill-in-the-blank and short answer questions, while alternative assessment questions involved students in correcting and interpreting illustrations, and drawing and explaining illustrations.

Bill and Laura looked at the responses from 110 students over a two-year period. They observed that students scored higher (x = 68.4% correct) on alternative assessment questions as compared to traditional assessment questions (x = 31.5% correct). They also noted that:

> Students who generally do well on a traditional style assessment showed more success on the traditional assessment questions. Those same students appeared to not do as well on the alternative assessment questions. Students who generally do poorly on a traditional style assessment showed greater success on the alternative style assessment questions. Those same students continued to do poorly on the traditional assessment questions.

Based upon the data collected during their action research project, Bill and Laura drew the following conclusion about alternative and traditional assessment questions:

> [There is a] need to use a combination of alternative and traditional style questions on assessments to reach a wide range of student learning styles. We do feel, however, that a higher concentration of alternative assessment questions tends to be more beneficial to the entire group of students than a higher concentration of traditional assessment questions.

The action research assignment enabled Bill and Laura to investigate an assessment issue of relevance to their classrooms, providing them with a classroom context for understanding assessment. The action research assignment gave meaning to alternative forms of assessment and provided empirical evidence for using alternative assessment questions in classrooms. Further, the action research assisted others in understanding the assessment issues, perspectives, and contexts facing Bill and Laura. We now know, for example, that an important issue for Bill and Laura is the ability of their students to perform successfully on summative assessments. We also know that alternative assessment for Bill and Laura includes the use of assessment questions such as correcting and interpreting illustrations and drawing and explaining illustrations.

TOOLS FOR THINKING ABOUT AND DEVELOPING ASSESSMENT TASKS

This section shares the tools that have been successfully used with teachers in the professional development process. These provide a guide for developing assessment tasks, a means for evaluating assessment tasks, and a resource for identifying what students should know and be able to do.

Based on the assessment literature (e.g., Herman, Aschbacher & Winters, 1992; Hymes, Chafin & Gonder, 1991; and Perrone, 1991), an assessment development and evaluation tool was developed and used to guide teacher development of assessment tasks as well as the evaluation of existing assessment tasks (Figure 2). To guide teachers in the development of multitrait assessment tasks, the following assessment task and scoring rubric template was developed (see Appendix E). The *Benchmarks for Science Literacy* (AAAS, 1993) was used as a tool for identifying the standards or proficiencies to be reached by students; that, in turn, guided the development of the assessment task and the scoring system. The benchmark is linked back to instruction to ensure "goodness of fit" or alignment between the curriculum and instructional activities.

Science for All Americans (AAAS, 1990) provided teachers with a tool that developed or enhanced their conceptual understanding of the science topic to be taught. Teachers used *Science for All Americans* as a resource for developing a concept map of the science topic. Teachers developed their concept map using Inspiration, Inspiration Software (1994). The concept map served as a curricular road map for the content to be taught, the sequence of coverage, and items to be assessed. Based on the concept map, teachers plan, sequence, and develop their science instruction. They search the science education research literature and the *Benchmarks for Science Literacy* (AAAS, 1993) to obtain information on students' ideas and conceptual understandings. This information about student conceptions was then used in planning and developing instruction and in developing the assessments to be used.

In the example provided, *Science for All Americans* (AAAS, 1990) and the *Benchmarks for Science Literacy* (AAAS, 1993) were utilized in developing the conceptual understanding, content knowledge, thinking processes, and science process standards for the instructional unit on earth processes (i.e., "The Processes that Shape the Earth" and "Habits of Mind"). The specific benchmarks for grades 3-5 used in identifying the performance standards were the following:

> Waves, wind, water, and ice shape and reshape the earth's land surface by eroding rock and soil in some areas and depositing them in other areas, sometimes in seasonal layers (AAAS, 1993, p. 73).

Developing Assessment Tasks
- Identify the instructional goals and the specific student performances (concepts, thinking processes, process skills, and behaviors).
- Indicate if the assessment task is to be used in a diagnostic, formative, or summative manner.
- Search the research literature for information on student conceptions, thinking and science process abilities and behaviors that will inform instruction and assessment task development.
- Search the literature for existing assessment tasks and/or identify potential tasks that would require students to demonstrate the performances. (Be cognizant of the instructional context).
- Identify assessment tasks that address multiple student traits or performances.
- State the specific criteria and standards for judging student performances.
- Identify assessment tasks that may be sequenced, producing an assessment system.
- Gather evidence to evaluate the assessment task and use results to refine assessment task and instruction.

Questions to Ask About Assessment Tasks
- Does the task align with curricular and instructional goals?
- Does the task fairly represent student abilities?
- Does the task enable students to demonstrate understandings, abilities, or progress?
- Does the task reflect authentic, real world situations?
- Does the task measure several student performances?

Recommendations for Developing Scoring System
- All important performance outcomes are represented by the criteria.
- Scoring system is easily interpreted and used.
- Criteria employ concrete references to performances and are clearly stated.
- Criteria reflect what the scientifically literate person should know and be able to do at the appropriate developmental level.
- Criteria are free of bias.
- Criteria reflect instruction and curriculum.

Figure 2. Suggestions for Designing Assessment Tasks

Keep a notebook that describes observations made, carefully distinguishes actual observations from ideas and speculations about what was observed, and is understandable weeks or months later (AAAS, 1993, p.293).

Make sketches to aid in explaining procedures or ideas (AAAS, 1993, p. 297).

Buttress their statements with facts found in books, articles, and databases, and identify
the sources used and expect others to do the same (AAAS, 1993, p.299).

Stream tables was selected as the assessment task because it aligned with the
instruction students had experienced during the "Earth Processes" unit. The
assessment prompt was then written to reflect the assessment task and the
performance criteria. The scoring rubric was developed with the assessment
performance stating what a scientifically literate fifth-grade student should know
and be able to do, a top performance level. Performance levels 2, 1, and 0 were
developed to differentiate student abilities. The assessment task and the scoring
rubric were then evaluated and revised based on student responses.

THE TECHNOLOGY OF ASSESSMENT

While others define performance assessment as involving students in the
manipulation of materials and equipment to generate a response, performance
assessment here refers to *any* assessment task in which students are required to
perform the assessment task. Characteristics of assessment tasks that were explored
and analyzed included: open-ended response tasks that provided students the
opportunity to generate a number of acceptable responses; closed-ended response
tasks where students provided a single best answer; and practical tasks where
students manipulated materials and used equipment to generate a response. All may
be used to assess inquiry and science process skills and conceptual understanding.

The development of assessment tasks involves: identifying the curricular and
instructional goals, stating the purpose of the assessment, identifying the task format
to be used, writing the task, developing the implementation procedures, constructing
the scoring system (guide or rubric), trial testing the task, analyzing student
responses, and revising the task. The first step in the development of an assessment
task is to identify the curricular and instructional goals that the task is to assess.
This ensures alignment between the curricular and instructional goals and the
assessment task. Further, it ensures that the assessment task addresses the science
domains: content knowledge, inquiry and science process skills, science attitudes,
personnel and cooperative skills, and thinking and reasoning skills. The second step
is to state the purpose of the assessment task. The purposes of assessment are many
and go beyond simply evaluating and grading students. The purpose of an
assessment task maybe to inform pedagogy, to diagnose student difficulties or
understandings, to measure student achievement or learning, or to provide feedback
to students.

After specifying the purpose of the task, the next step is to identify the
assessment format to be used. The format will in part depend on the purpose of the
assessment. For example, if the purpose is to assess students' inquiry and process
skills in a laboratory setting, then the format of the assessment is likely to involve a
practical task. In determining the task format it is necessary to consider whether
individual students, pairs of students, or student groups, will complete the task. Will
the task involve an extended investigation or an on demand task? Will it assess
multiple student traits or science domains? The task format selected should be such

that the assessment system reflects a variety of task formats and provide multiple measures.

Depending on the task format, the task may contain the following sections: the setting or context, the prompt, and directions for completing the task. The task design comprises these three dimensions. The setting or context provides students with the necessary background information for understanding and completing the task. The prompt states what students are to do to complete the task, guiding students in their thinking and actions. Directions provide students with additional information about how to complete the task, what materials or equipment they can use, or how they are to respond to the prompt. The task may require students to respond by writing a short brief, constructing a model, creating a PowerPoint presentation, or making an oral presentation. These task dimensions need to be considered in writing the task as well as the science domains or student traits to be assessed. The task context, prompt, and directions must all be aligned with the science domains to be assessed or the task will not reliably measure what is intended.

The structure of the scoring system is dependent on the task format and the task design. The task design influences what students do and the science domains or student traits exhibited. This means that the scoring system must be consistent with the task design. Scoring rubrics may be written as analytical or holistic rubrics and as either generalized or specific in stated criteria. Holistic scoring rubrics contain multiple levels of student performance. Within each level, multiple categories or criteria for student performance are described. The criteria for the top performance level should reflect state or national standards that indicate what the scientifically literate student should know and be able to do. Analytic rubrics consist of a series of individual statements, or criteria, used to assess individual components of student performance. The decision to write an analytic or holistic rubric should be based on whether the student responses are to be assessed as a complete product or in subcomponents. Although generalized scoring rubrics are easier to write, they lack criteria specific to the task. Because the criteria should be developed in alignment with state and national standards, it is also helpful to write a response that would reflect a scientifically literate student. This hypothetical student response may be used to operationalize a top performance level for a holistic rubric, or may be broken down into its subcomponents to provide the individual criteria for the analytic rubric. Another means to develop a scoring rubric is to identify the performance categories and then describe the expected indicators for each performance category. The expected indicators are then written into the form of either an analytic or holistic scoring rubric.

The implementation procedure describes the steps to prepare the task for use and how the task should be implemented. The procedure will depend on the task format. It should be clear and concise that others could successfully implement the task. Issues to consider include time to complete the task, special materials for completing the task, safety concerns, special needs students, and possible sequencing of assessment tasks. After the assessment task is developed it must be trial-tested to ensure its quality. Trial-testing need not be done with every student, but only with a few who are representative of the academic, gender, and ethnic makeup of the classroom.   Trial-testing involves analyzing student responses and obtaining feedback from students concerning the task. The task is revised based on the student responses and feedback. Student work samples should be kept as evidence of the task's quality and as a record of student performance. The task may need further revision based on subsequent use with different students.

TEACHERS AS CONSUMERS OF ASSESSMENT

Not only do teachers need to become educated developers of assessment, but they must also become educated consumers of assessment. As such, teachers must become knowledgeable about the characteristics of quality assessments. Herman, Aschbacher and Winters (1992) identified key questions to guide the evaluation of assessments:

- Does the assessment match instruction?
- Does the assessment adequately represent the content and skills taught?
- Does the assessment enable students the opportunity to demonstrate progress?
- Is the assessment authentic?
- Is the assessment multitrait?

In addition to these guiding questions, the Center for Research on Evaluation, Standards, and Student Testing (CREST) have identified criteria for determining the quality of assessments (Linn, Baker, & Dunbar, 1991): consequences, fairness, transfer and generalizability, cognitive complexity, content quality, content coverage, meaningfulness, and cost and efficiency. A neglected aspect of evaluating and using existing assessment tasks is the consideration of the instructional context of the assessment task. Adopting an existing assessment without considering its alignment with classroom instruction or modifying the assessment to align with instruction may reduce the reliability and validity of the assessment.

To give teachers experience as assessment consumers, participants were provided with examples of existing assessments, both teacher and professionally developed, to review and critique. Teachers as consumers of assessment must also learn to be critical of their own and colleague developed assessments. Experience shows that teachers often are not significantly critical of the assessment tasks they develop and tend not to be critical of colleagues' assessments, likely in an attempt to maintain trust and community. Teachers, however, are critical of professionally developed assessments.

CONCLUDING THOUGHTS

This professional development framework was built upon the establishment of an intellectual community of learners in which participants were viewed as professionals and where they engaged in critical reflection, collaborative dialogue, and the negotiation of a shared understanding of teaching and assessment. The professional development framework took into account the school culture and the classroom community where participating teachers practiced. The professional development process established a community in which teachers' understandings about practice were shared and challenged through reflection and collaborative dialogue, and where alternatives to practice were experienced, leading to change.

The professional development framework provides a successful model for operationalizing the NRC (1996) standards: engaging participants in intellectual activities concerning practice, promoting collaborative dialogue about practice, supporting the collaborative development of instructional and assessment materials, and connecting the professional development process to the context of the teachers' school and classroom. The process incorporated the NRC (1996) professional development standards by engaging participants in learning about assessment through activities that involved students and utilized real student work as the basis for thoughtful reflection, interaction with colleagues, and application to practice. Participants also observed good classroom assessment practice. They reviewed assessment tasks, aligned curriculum, instruction, and assessment, selected and developed assessment tasks; analyzed and interpreted assessment data; and collaborated with others to evaluate student work, and to develop and refine assessment tasks (NRC, 1996).

ACKNOWLEDGEMENTS

Portions of this chapter are based on work supported by the National Science Foundation (Grant No. TPE-9154840). Any opinions, findings, and conclusions or recommendations expressed in this chapter are those of the author and do not necessarily reflect the views of the NSF.

REFERENCES

American Association for the Advancement of Science (1990). *Science for All Americans*. New York: Oxford University Press.

American Association for the Advancement of Science (1993). *Benchmarks for Science Literacy*. New York: Oxford University Press.

American Association for the Advancement of Science (1998). *Blueprints for reform: Science, mathematics, and technology education*. New York: Oxford University Press.

Darling-Hammond, L., Ancess, J., & Falk, B (1995). *Authentic assessment in action: Studies of schools and student work*. New York: Teachers College Press.

Doyle, W. & Ponder, G.A. (1977). The practicality ethic in teacher decision-making. *Interchange, 8*, 1-12.

Eisner, E.W. (1985). *The educational imagination: On the design and evaluation of educational programs* (2nd ed.). New York: Macmillan.

Eisner, E.W. (1994). *Cognition and curriculum reconsidered* (2nd ed.). New York: Teachers College Press.

Feely, J. (1986). *Reading practices in schools: Linking theory with practice*. (ERIC Document Reproduction Service No., ED 269 734.)

Fenstermacher, G. D. (1979). A philosophical consideration of recent research on teacher effectiveness. *Review of Research in Education, 6*, 157-185.

Fenstermacher, G. D. (1986). Philosophy of research on teaching: Three aspects. In M.C. Wittrock (Ed.), *Handbook of research on teaching*. New York: Macmillian.

Fullan, M. (1990). Staff development, innovation and instructional development. In B. Joyce (Ed.), *Changing school culture through staff development*. Alexandria, VA: Association for Supervision and Curriculum Development.

Herman, J.L., Aschbacher, P.R., & Winters, L. (1992). *A practical guide to alternative assessment*. Alexandria, VA: Association for Supervision and Curriculum Development.

Huberman, M. (1991). Teacher development and instructional mastery. In A. Hargreaves & M. Fullan (Eds.), *Understanding teacher development*. London: Cassells.

Huberman, M. (1995). Networks that alter teaching: Conceptualizations, exchanges and experiments. *Teachers and Teaching: Theory and Practice*, 1(2), 193-211.

Hymes, D.L., Chafin, A.E., & Gonder, P. (1991). *The changing face of testing and assessment: Problems and solutions*. Alexandria, VA: Association for Supervision and Curriculum Development.

Inspiration Software (1994). Inspiration for windows. Portland OR: Inspiration Software, Inc.

Linn, R.L., Baker, E.L., & Dunbar, S.B. (1991). Complex, performance-based assessment: Expectations and validation criteria. *Educational Researcher*, 20, 15-23.

McNiff, J. (1993). *Teaching as learning: An action research approach*. London: Routledge.

Nespor, J. (1987). The role of beliefs in the practice of teaching. *Journal of Curriculum Studies*, 19, 317-328.

National Research Council. (1996). *National science education standards*. Washington, DC: National Academy Press.

Perrone, V. (1991). *Expanding student assessment*. Alexandria, VA: Association for Supervision and Curriculum Development.

Perry-Sheldon, B., & Allain, V.A. (1987). *Using educational research in the classroom*, Bloomington, IN: Phi Delta Kappa Educational Foundation.

Olson, J. (1980). Teacher constructs and curriculum change. *Journal of Curriculum Studies*, 12, 1-11.

Olson, J. (1981). Teacher influence in the classroom: A context for understanding curriculum translation. *Instructional Science*, 10, 259-275.

Richardson, V. (1990). Significant and worthwhile change in teaching practice. *Educational Researcher*, 19, 10-18.

Robottom, T.M. (1987). Two paradigms of professional development in environmental education. *The Environmentalist*, 7, 291-298.

Sarason, S.B. (1971). *The culture of school and the problem of change*. Boston, MA: Allyn and Bacon.

Simmons, J. (1984). *Action research as a means of professionalizing staff development for classroom teachers and school staff*. (ERIC Document Reproduction Service No., ED 275 639.)

Shulman, L.S. (1987). Knowledge and teaching: Foundations of the new reform. *Harvard Educational Review*, 57, 1-22.

Taylor, P. (1993). Collaborating to reconstruct teaching: The influence of researcher beliefs. In K. Tobin (Ed.), *The practice of constructivism in science education*. Washington, DC: AAAS Press.

Zahorik, J. (1984). Can teachers adopt research findings? *Journal of Teacher Education*, 35, 34-36.

APPENDIX A

REFLECTION ON CURRENT PRACTICE:

GUIDELINES AND PROCEDURES

Purpose

The literature suggests that changes in practice occur when practitioners have opportunities to reflect upon their current practice. These kinds of opportunities enable practitioners to construct a profile of their current practice, to analyze data from their own classrooms, to discuss their problems and solutions with colleagues, and to discuss their successes and failures. The purpose of this activity is to assist you in reflecting upon your current instruction and assessment practices.

Materials
You should have received the following materials to assist you in your self-reflection: 2 videotapes, notebook and a sealed envelope.

Procedures
You will develop a profile of an instructional unit, including the assessment techniques, for one class. Select an instructional unit that you consider to be typical. To develop your profile you will need to do the following:

1. Videotape the instruction from the beginning to the end. If you need to use additional videotapes you will be compensated at a later date.

   a. If possible, have another individual run the video camera. The individual should focus the camera so that you are at the center, including as much of the class as possible. If an individual is not available to operate the video camera, station the camera so that you are at the center and include as much of the class as possible. Depending upon your situation, you may have to move the camera to accommodate changes in your location.

   b. If you have a remote microphone available, please wear the mic to improve sound quality.

   c. Record your name, grade, date, and unit content on the videotape.

2. Keep a personal log of the instructional unit from the start to the end. Record your reflections about the instructional unit for each day. To assist your reflection you may wish to address these questions: What happened? What did I do? What did my students do? How do I feel about what happened? What did I want to happen? What problems occurred? What success occurred? Be sure to indicate the date and time for each entry.

3. Keep a copy of everything you provide to students during the unit--worksheets, handouts, lab sheets, and tests. You will also need to duplicate the textbook chapter, laboratory manual and lesson plans related to the instructional unit.

4. After you have completed the instructional unit, open the sealed envelope and answer the questions and follow the procedures for using your videotapes to reflect upon your instruction. It is important that you not open the sealed envelope until you have completed the instructional unit. The reason for not peeking is to prevent a bias in your current practice. The self-reflection only has meaning and value when based on your current practice.

5. Select a segment of your videotape that you wish to share and discuss with a small group of colleagues at our next meeting. Also be prepared to discuss your instructional and assessment profile.

APPENDIX B
ALTERNATIVE ASSESSMENT ASSIGNMENT

Introduction/Purpose

Science educators are being called upon to reform their assessment practice and to align assessment with their science curriculum and instructional approaches. The successful implementation of innovations in science teaching rests on the development and use of appropriate assessment techniques, which are connected to the curricular and instructional innovation. The purpose of this assignment is to engage you in developing, implementing, and reflecting upon alternative assessment techniques.

Procedures

You will develop, implement, and reflect upon two different alternative assessment techniques. To accomplish this you will need to:

1. Select two assessment instruments that you currently use and will use to evaluate your students between now and the end of school.

2. Develop one alternative assessment prototype for each of the above assessment instruments. This means that you will develop two different alternative assessment prototypes. The alternative assessment prototypes need not assess every item covered on your original (current) assessment instruments. As you develop your alternative assessment prototypes be sure to include the following:

- Student directions for completing the alternative assessment. Are the directions clear and concise? Is all necessary information provided?
- Scoring system, which is used to judge student performance. Does the scoring system reflect a range in student performance? Does the scoring system state specific criteria for each level of performance?
- Provide all necessary information for administering the alternative assessment prototypes to students; that is, will other teachers be able to use your prototypes?

Note: This is a first attempt and you will need to revise your prototypes based on student responses. Alternative assessment techniques must be revised based on student responses in order to improve their reliability and validity.

3. Administer both your current assessment instruments and your alternative assessment prototypes to at least one class. Depending on the length of your assessment instruments, you may want to administer your current assessment instruments on one day and your alternative assessment prototypes the next day.

4. Photocopy the student responses for all assessment instruments (current and alternative assessment prototypes), along with clean copies of each assessment instrument. You will need to turn in the photocopies with your reflection paper.

5. Based on the results for one class, write a reflection paper that compares and contrasts student outcomes in your current assessment instruments and the alternative assessment prototypes. The following questions may assist you in your reflection:

- Did students exhibit a difference in their understanding? If so, in what ways? Why might this be? What is your evidence?
- In your way of understanding, which instrument (current or alternative) best reflected student learning? Why do you think so? What is your evidence?
- What are the benefits and difficulties of developing and implementing alternative assessment techniques? Give reasons for your views.
- Based on student responses, in what ways would you modify your prototypes to improve their reliability and validity?

6. Revise your prototypes based on student responses and your reflection. Be sure to photocopy and turn in your revised prototypes.

7. Prepare a short (10-15 min.) small group presentation on your reflection paper. You should at least share examples of your alternative assessment prototypes, student evidence (results), and your general thoughts about the benefits and difficulties of alternative assessment techniques.

APPENDIX C
INTEGRATING INSTRUCTION AND ASSESSMENT ASSIGNMENT

Introduction/Purpose
Science educators are being called upon to reform their assessment and to align assessment with their science curriculum and instructional approaches. The successful implementation of innovations in science teaching rests on the development and use of appropriate assessment techniques, which are connected to the curricular and instructional innovation. The purpose of this assignment is to engage you in developing, implementing, and reflecting upon assessments aligned with instruction.

Procedures
You will develop, implement, and reflect upon an instructional sequence containing assessment tasks that are integrated into instruction as diagnostic, formative, and summative. To accomplish this you will need to:

1. Develop the assessment tasks.   One diagnostic, one formative, and one summative that align with your instructional sequence. This means you will develop at least three different assessment tasks. Be sure to include the following:

- Student directions for completing the assessment. Are the directions clear and concise? Is all necessary information provided?
- Scoring system used to judge student performance. Does the scoring system reflect a range in student performance? Does the scoring system state specific criteria for each level of performance?
- Provide all necessary information for administering the assessment tasks to students; that is, will other teachers be able to use your assessments?

Note: Assessment tasks must be revised based on student responses in order to improve their reliability and validity.

2. Videotape and collect student work examples from at least one class to reflect on and to share with participants at our next workshop.

3. Photocopy the student responses for all assessments, along with clean copies of each assessment. You will need to turn in the photocopies with your reflection paper.

4. Based on the results for one class, write a reflection paper that anaiyzes student performance, addressing the alignment between instruction and assessment. The following questions may assist you in your reflection:

- Did students exhibit a difference in their understandings? If so, in what ways? Why might this be? What is your evidence?
- What are the benefits and difficulties with developing and implementing assessments as diagnostic, formative, and summative? Give reasons for your views.
- Based on student responses, in what ways would you modify your assessment to improve reliability and validity? What ways might you modify your instruction?
- How did each of your assessments align with your instruction?

5. Revise your assessment tasks and instruction based on student responses and your reflection. Be sure to photocopy and turn in your revised assessments and instructional sequence.

6. Prepare a short (10-15 min.) small group presentation on your reflection paper. Your presentation should include a 3-5 min. video clip from your classroom. You should at least share examples of your instructional activities, assessment tasks, student work, and your general thoughts about the benefits and difficulties of implementing assessment as diagnostic, formative, and summative.

APPENDIX D

AUTHENTIC ASSESSMENT ACTIVITY: STRUCTURE OF MATTER

Developed by: Paul Adams and Dan Shepardson

A. Overview

Students will observe the making of fudge either at home, in the classroom, or on videotape. Students will be asked to make a narrated pictorial that represents the changes that occur to the ingredients over time. In addition to identifying the states of matter the students will be asked to indicate the energy transfer occurring in the mixture.

B. Fudge Directions

A common recipe adapted from Kraft Marshmallow Crème, *Easy Fantasy Fudge*:

Preparation Time: 10 minutes plus cooling
Cooking Time: 15 minutes
1 1/2 cups (3 sticks) margarine
1 jar (13 ounces) marshmallow crème
6 cups sugar
1 1/3 cups evaporated milk
2 teaspoons vanilla
2 packages (12 ounces each) semi-sweet real chocolate chips
1 chocolate bar

Mix margarine, sugar and milk in heavy 5-quart saucepan; bring to full boil, stirring constantly. Continue boiling 5 minutes over medium heat, or until candy thermometer reaches 234 F, stirring constantly to prevent scorching. Remove from heat. Gradually stir in chips and broken chocolate bar until melted. Add remaining ingredients; mix well. Pour into 3 greased 9-inch square or 2 greased 13 x 9-inch baking pans. Cool at room temperature; cut into squares. Makes about 6 pounds.

C. Student Directions and Data Collection

On the response sheet use words and pictures to describe the states of matter of the ingredients (fudge stuff), the energy transfer occurring, the changes to the states of matter that are caused by the energy transfer, and the evidence to support your answers. Be sure to respond to each of the four steps involved in making the fudge. The four steps are:

1. Mix margarine, sugar and milk, bring to a boil.
2. Continue boiling for 5 minutes.
3. Stir in candy and other ingredients.
4. Pour into pans.

D. Scoring Rubric Template

Note: These scoring rubrics are to be completed for each of the four sheets. Specific observations and documenting evidence needs to be developed within the context of each classroom situation. Keep in mind that the scoring rubrics provide a template of what to look for when assessing the writing and pictorial information that the students are providing.

States of Matter

Points	Criteria
3	Notes original state and final state of ingredients in a given step based upon the classification scheme (e.g., solid, liquid, gas).
2	Notes original state and final state of ingredients in a given step based upon the classification scheme for at least 50% of the ingredients.
1	Notes original state and final state of ingredients in a given step based upon the classification scheme for some of the ingredients.
0	No work or inconsistent classifications.

Evidence for States of Matter

Points	Criteria
2	Explains all inferences based on identified physical properties.
1	Provides some explanation for inferences based on identified physical properties.
0	Provides no consistent explanation for inferences

Energy Transfer

Points	Criteria
3	Notes how energy is transferred into or out of the mixture in terms of the changes of state of the ingredients in a given step.
2	Notes how energy is transferred into or out of the mixture in terms of the changes of state of the ingredients in a given step for at least 50% of the ingredients.
1	Notes how energy is transferred into or out of the mixture in terms of the changes of state of the ingredients in a given step for some of the ingredients.
0	No work or inconsistent observations.

Evidence for Energy Transfer

Points	Criteria
2	Explains all changes resulting from energy transfer in terms of physical observables (e.g. melting, release of vapor, hot to touch, rising thermometer, burner on stove in contact with pan).

1   Explains some of the changes resulting from energy transfer in terms of physical observables (e.g. melting, release of vapor, hot to touch, rising thermometer, burner on stove in contact with pan).

0   Provides no consistent explanation for how changes are related to energy transfer.

APPENDIX E

ASSESSMENT TASK AND SCORING RUBRIC TEMPLATE

The prompt: Using the equipment and materials at your station, design and conduct a stream table investigation for simulating how moving water and melting ice effect the earth's surface. Record your investigation design, data collected, and observations in your journal. Using the data and observations recorded in your journal and the resource materials provided, write a geological report explaining your results.	
Description of performances to be assessed in the assessment task (Based on the Benchmarks)	Description of performances demonstrated in the assessment task
Conceptual understanding: Water and ice shape and reshape the earth's land surface by eroding rock and soil in some areas and depositing them in other areas.	
Content or factual knowledge: Erosion, sediment, deposition, slope, soil, sand, delta, glacier, and glacial till.	
Thinking/reasoning processes: Buttresses statements with facts found in books and CD-ROM, and identify the sources used.	
Science processes: Keeps a notebook that describes observations made, carefully distinguishes actual observations from ideas and speculations about what was observed. Makes sketches to aid in explaining procedures or ideas.	

Holistic Rubric for Assessment Task			
Conceptual Understanding	Content Knowledge	Thinking Processes	Science Processes
Performance Level 3 Explains how water and ice shape and reshape the earth's land surface by eroding soil and sand in some areas and depositing them in other areas.	Accurately applies 8 of the following concepts in explaining the results of the stream table activity: erosion, sediment, deposition, slope, soil, sand, delta, glacier, and glacial till.	Statements and explanations are supported by reference to the textbook and CD-ROM.	Notebook describes observations of stream table, distinguishes actual observations from ideas about what was observed. Sketches relate to and aid in explaining how water and ice reshape the earth's land surface.
Performance Level 2 Explains only how water or ice shape and reshape the earth's land surface by eroding soil and sand in some areas and depositing them in other areas.	Accurately applies 5 to 7 of the following concepts in explaining the results of the stream table activity: erosion, sediment, deposition, slope, soil, sand, delta, glacier, and glacial till.	Statements and explanations are partially supported by referencing the textbook or the CDROM, but not both.	Notebook describes some observations made of stream table, distinguishes actual observations from ideas about what was observed. Some sketches relate to and aid in explaining how water and ice reshape the earth's land surface.
Performance Level 1 Indicates only that water and/or ice can change (shape and reshape) the earth's land surface. Fails to explain how water and ice erode soil and sand in some	Accurately applies 4 of the following concepts in explaining the results of the stream table activity: erosion, sediment, deposition, slope,	Statements and explanations are made, but are not supported.	Notebook describes some observations made of stream table, dose not distinguishes actual observations from ideas about what was observed. Sketches do not

areas and deposits them in other areas.	soil, sand, delta, glacier, and glacial till.		relate to and aid in explaining how water and ice reshape the earth's land surface.
Performance Level 0 Not attempted	Not attempted	Not attempted	Not attempted

EDITH S. GUMMER AND DANIEL P. SHEPARDSON

3. THE NRC STANDARDS AS A TOOL IN THE PROFESSIONAL DEVELOPMENT OF SCIENCE TEACHERS ASSESSMENT KNOWLEDGE AND PRACTICE

Assessment has become a central topic in recent discussions of educational reform. The movement toward discipline-based standards emphasizes the need for the development of new and improved assessments that are closely tied to changes in curriculum and instruction at local, state, and national levels. Standards-based classroom assessment expands the evidence gathered about what students know and can do in science. Such assessment should be guided in its development by state and national standards documents, and be linked to the National Research Council [NRC] (1996) *National Science Education Standards*. The NRC Standards specifically address the nature of the assessment processes both inside and outside the local science classroom.

The complex nature of assessment requires that professional development programs address a number of issues that have multiple interconnections. These issues include introducing the: 1) national and state science education standards, 2) purposes, formats, and characteristics of good science assessment, 3) relationship between science content, inquiry, reasoning, and thinking processes assessed in such assessments, 4) development of scoring rubrics that capture the intended performance domains, and 5) psychometric characteristics of the assessment tasks.

This chapter describes the professional development activities that assisted teachers explore ways to expand their assessment practice. It includes an example of how "assessment" was introduced to participants and an example of an assessment task that was used as a model of "alternative" assessment. This is followed by a discussion of the activities that engaged participants in an exploration of the state and national science education standards. The development of criteria to judge assessments and assessment practices will constitute a major part of that discussion. Finally, we discuss other assessment issues related to the professional development of teachers in science.

INTRODUCING ASSESSMENT TO TEACHERS: MOVING TOWARD A CONSENSUAL UNDERSTANDING

Daniel P. Shepardson (ed.), Assessment in Science, 39—51.

Professionals organize their understandings, knowledge, and skills into episodes, events, cases, or stories (Bruner, 1988; Carter & Doyle, 1989). They also contain descriptions of the physical setting, the activities that were occurring, what the students were doing, what the teacher was doing, and what the classroom was like. In this project, teacher descriptions or stories about classroom practice and their reflections on their stories provided a simple way of introducing and addressing issues of importance to classroom practice (Frederick, 1990; Schon, 1983). To engage teachers in reaching a consensual understanding of assessment, participants wrote a two to three-page description of their ideal classroom as a pre-professional development activity. The teachers' descriptions of the ideal practice reflected their understandings of classroom practice; these were teacher stories about teaching, learning, and assessing. These stories were then shared in small groups as the participants were required to look for what they saw as "assessment" embedded in them or opportunities for incorporating assessment. From this activity the nature of the term "assessment" and the lack of evidence of the same in the participants' stories began to develop.

It was important to introduce the notion of standards early in the professional development program; therefore, a second activity introduced the *National Science Education Standards* (NRC, 1996). A summary of the standards was provided as an overview, especially of the assessment standards. In small groups, participants reflected upon the match between the NRC Standards and their ideal classroom descriptions or stories. Participants quickly realized the lack of attention to assessment in their ideal classroom descriptions.

These two activities allowed participants to put forth their ideas and understandings about classroom practice, providing a means to compare actual teaching practice with the national science education standards. By placing the activities in the context of "ideal" practice it removes the activity from one of evaluating teacher practice to reflecting on and thinking about practice. Placing the activity in the context of story provided a means to process information (Hardy, 1977) and enabled teachers to explore future alternatives to current practice (Kazemek, 1985).

Next an assessment task was modeled to provide a context for teachers to reflect on alternative assessment. This activity successfully used an inquiry- and laboratory-based assessment task developed by teachers in New York (Figure 1). It required participants to design and carry out an experiment to verify a claim put forward by the distributor of a diet supplement. This was not intended as an example of a "perfect" assessment, but instead emphasized important assessment questions that then became the topics of subsequent discussion. The participants constructed and evaluated their own individual responses to the task and then considered the prior knowledge required for students to complete the task in terms of the concepts and thinking and inquiry processes embedded in the task.

From this experience, participants determined that students would need to understand the nature of the activity of an enzyme and to recognize that glucose and galactose are the product of lactose digestion. Students also needed to understand and set up a controlled experiment. To conduct the controlled experiment students needed to indicate that they would test the glucose in one well spot with a test tape to determine the nature of the indicator of glucose. Students also needed to set up two well spots, one with Lactaid and water and one with Lactaid and milk. These served as controls containing only the substrate with water or the enzyme with water. A third well spot was the experimental well.

Lactaid

Assessment task designed by:
Patricia Barker, Newburgh Free Academy/OCCC
Mary Jean Bahret, Newburgh Free Academy
Sue Holt, Williamsville East High School

Many people after eating dairy products such as milk or ice cream experience digestive discomforts in the form of flatulence (gas), bloating, cramping and even diarrhea. African-American, Asian, Hispanic, and Native American populations have a higher percentage of people who experience discomfort. Doctors believe that these individuals fail to produce the enzyme, lactase, and do not readily digest lactose to galactose and glucose. This disorder can be inherited. The product Lactaid claims to help these people by supplying the enzyme. Does Lactaid work?

The box that contains Lactaid makes these statements:

What is Lactaid? Lactaid, the original dairy digestive supplement, is a simple and natural way to make milk more digestible.
How does Lactaid Work? Add 5-7 drops to a quart of milk, shake gently and refrigerate for 24 hours."

Using only the included materials, design and carry out a safe scientific experiment to determine if Lactaid could help lactose intolerant people. You must wear goggles while performing your experiment.

Before beginning, make sure that your kit contains the following. Place an "a" if present.

___distilled water	___glucose solution in beral	___glucose test strips
___lactaid in beral	___mile in beral	___paper towel
___plastic stirrers	___Q-tips for clean up	___24-well plate

NOTE: Please make sure the unused glucose test strips are re-wrapped in foil to prevent exposure to light and humidity. Also reassemble the kit for use by the next student.

Figure 1. Sample Alternative Assessment Task

Several design questions arose about what the assessment task communicates to students. Participants questioned whether the task could be answered without any particular knowledge of biology given the information in the prompt.   Other participants argued that there was a great deal of biological understanding involved concerning the nature of enzymes required to complete the task. Several participants were concerned about the nature of the level of communication necessary to successfully respond to the task. These participants wanted to provide students with more structure about how to communicate their results. This demonstrated the editorial aspect of assessment development, where almost everyone asked to comment on a developed assessment task have constructive comments that modify the task.

The participants then considered the "Changing Emphases" section of the Assessment Standards chapter in the *National Science Education Standards* (NRC, 1996, p. 100) shown in Table 1. They determined that the task measured highly valued experimental design skills and concepts central to biology, but they were undecided about whether the knowledge measured was discrete or rich; although they came to no consensus about this issue, they agreed that the task measured scientific understanding and reasoning rather than just scientific knowledge. They could not agree on whether the task measured what students *did* know rather than what students *did not* know. They decided that the task as presented gave no indication about the opportunity to learn. This, they argued, would require some indicator of what experiences the teacher had provided for the students.   The participants determined that the task represented a discrete assessment by the teacher rather than an ongoing assessment by students. Although teachers had developed the task, only its use in assessment external to the classroom would fulfill the final changed emphasis.

Table 1. The Changing Emphasis of Assessment

Less Emphasis On	More Emphasis On
Assessing what is easily measured	Assessing what is most highly valued
Assessing discrete knowledge	Assessing rich, well-structured knowledge
Assessing scientific knowledge	Assessing scientific understanding and reasoning
Assessing to learn what students do not Know	Assessing to learn what students do understand
Assessing only achievement	Assessing achievement and opportunity to learn
End of term assessments by teachers	
Development of external assessments by Measurement experts alone	Students engaged in ongoing assessment of their work and that of others
	Teachers involved in the development of external assessments

Based on the *National Science Education Standards* (NRC, 1996, p. 100).

Participants were then asked to consider several aspects of assessment tasks that were presented (see the Matrix of Classroom Assessment Practice, Chapter 2).

Although these assessment task dimensions overlap with the "Changing Emphases," they were introduced to the participants to initiate their thinking about the aspects of assessment task design and assessment practice. Participants were asked to think of the dimensions as anchoring opposite ends of several continua that described the dimensions of assessment practice.

Participants rated the "Lactaid Task" as having an on-demand nature; in other words, the students were required to finish the task in a single session as prompted by the teacher. The task was determined to be an independent activity because no group involvement was indicated. The participants could find no evidence that the work of students was supported by either the teacher or by outside materials. Students' responses to the task were to be written, and the nature of that communication became the subject of some discussion. The participants determined that the task had one "correct" answer although the format of that answer could be relatively open-ended. Participants were enthusiastic about the application of skills and knowledge required by the task, but several again argued that the task could be performed with only the information in the prompt and without additional understanding of biology.

The notion of "performance" required significant interpretation by the participants. They argued that even the response to a multiple-choice question involved a performance of sorts. Although students had to actually carry out a controlled experiment, participants felt that the structure of the task was somewhat contrived. While these elicited discussion, they did not help the teachers as they thought about the development of assessment tasks. They felt that the requirements of experimental design and implementation represented the process of inquiry within the science classroom. Finally, they determined that the task represented only the discipline of biology.

Next teams of participants were asked to consider the purposes for which assessments might be constructed. They discussed the different types of decisions that assessment data might inform at the classroom level, including information for teachers and students. The discussion expanded to address ways in which teachers and administrators, parents and the community, and state agencies might use assessment data.

Finally, the group engaged in brainstorming alternative assessment formats. This led to a discussion of how a teacher might go about changing his or her assessment practice. This led to the first reflection assignment in which teachers considered how to modify an existing assessment (in this case, a series of multiple-choice items) to get a broader picture of what students know or can do. As a part of the first assignment, participants were requested to put together a "snapshot" or portfolio of their current assessment practices. This picture was used as an introductory activity in the next professional development workshop. An example is provided below. This is followed by a set of definitions that summarize teachers' understandings about various assessment concepts.

Assessment Format	Percentage of Practice
Teacher constructed quizzes	10
Teacher constructed tests	18

In-class worksheets	15
Laboratory reports	40
Oral reports	2
Short written reports	5
Long written reports	10

Assessment: Any means of measuring a student's class work, knowledge, understanding and/or ability to use that learned information. Assessment can also be used to evaluate many other aspects of the class, including but not limited, to curriculum and teacher or project effectiveness.

Testing: Another term describing assessment.

Evaluation: Any means of measuring a stated objective.

Alternative Assessment: A nontraditional assessment form.

Authentic Assessment: An assessment that is nontraditional and relates or applies the learning activity, concepts, or goals to a real life situation.

Performance Assessment: An assessment that is based on an activity in which students demonstrate knowledge or understanding.

In summary, the first workshop experience in the professional development program was designed to prepare teachers to think about assessment practice, assessment task design, assessment issues, and the state and national science education standards. This process occurred through the active involvement of teachers, engaging them in an intellectual process of reflecting on assessment. This was the building block for future professional development activities.

NATIONAL STANDARDS AS TOOLS FOR UNDERSTANDING ASSESSMENT

The next step in the professional development program was to bring the issue of science education standards to the foreground. Because participants often had varied backgrounds and experiences, it was essential that a consensual understanding of the *National Science Education Standards* (NRC, 1996) be developed. To facilitate discussion and consensus building, an assessment inventory was administered to elicit participants' definitions of "standards" as well as other assessment constructs. These then served as the starting point for discussion. In general, two types of definitions emerged from teacher responses; first, "standards" was viewed as examples of high quality referents to which other objects or practices might be compared. Second, "standards" were seen as hurdles for students to overcome in order to proceed or succeed. Participants debated the nature of these two definitions and concluded that they viewed standards primarily as indicators of high quality science education.

Participants then researched the definition of "standards" in the *National Science Education Standards* (NRC, 1996). Participants read through the introduction to the document and found that "standards" were described as:

....criteria to judge quality: the quality of what students know and are able to do; the quality of the science programs that provide the opportunity for students to learn science; the quality of science teaching; the quality of the system that supports science teachers and programs; and the quality of assessment practices and policies. Science education standards provide criteria to judge progress to a national vision of learning and teaching science in a system that promotes excellence, providing a banner around which reformers can rally (p. 12).

The *Benchmarks for Science Literacy* (American Association for the Advancement of Science [AAAS], 1993) was used as a secondary "standards" document. The *Benchmarks* provided participants with background information and was used in developing common definitions of "scientific literacy" and "science for all." Based on the *Benchmarks*, the following dimensions of scientific literacy may be identified:

- There are different levels of knowing and habits of mind based on grade or developmental level.
- Involves an understanding of the interrelationship among science, mathematics, and technology.
- That a scientifically literate person may not be able to do science, but that they are able to use knowledge and habits of mind to think about and make sense of science.
- Enables a person to observe events and think about such events, as well as to understand explanations of the observed events.
- That a person has the knowledge and habits of mind to make decisions and take actions.

"Science for all" was viewed as the idea that no individual or group should be excluded from an opportunity to become scientifically literate. Teachers felt they could provide an opportunity for all students to learn science, but that it is difficult to ensure that all students become scientifically literate. Teachers also perceived the *Benchmarks* as indicators of student performance: the standards or levels of performance that students should attain.

Selected readings from *National Science Education Standards* (NRC, 1996) and the *Benchmarks for Science Literacy* (AAAS, 1993) were used to explore how a teacher might gather evidence to answer the following questions:

- Has the teacher provided students with the opportunity to use their imaginations to solve science problems and analyze the results?
- Were students given the opportunity to communicate their findings and receive constructive, critical feedback from their peers?
- Were students provided with opportunity to examine various points of view and determine possible sources of bias?

These questions assisted participants to reflect on the relationship between the criteria for assessment they had developed from the *National Science Education Standards* (NRC, 1996) and the *Benchmarks for Science Literacy* (AAAS, 1993).

They also helped expand the teachers' understandings of the criteria they would use to evaluate their assessment tasks and practice.

The next professional development activity required teachers to identify how standards applied to assessment and to determine how they might use the notion of "standards" to develop and implement quality assessments. Participants individually read the Assessment Standards chapter in the *National Science Education Standards* (NRC, 1996) and wrote a response to the following question: "What criteria for assessment in science education do the NRC Standards recommend to judge ideal teacher practice?"

Participants then discussed the nature of this question, clarifying the concepts of "ideal assessment tasks" and "ideal teacher assessment practice." Teachers were encouraged to recognize that both the assessment tasks and the assessment practices that surrounded those tasks should be available for evaluation. Participants then came together in small groups to develop lists of such criteria to present to the larger group. An example of the list produced by one group is shown in Table 2. As the groups compared the criteria they developed, they began to see overlap and repetition.

The groups then used these criteria to evaluate their own classroom practices. It became apparent that many felt they were not effectively assessing either their students or their own instructional and curricular decisions. The teachers became concerned that any particular snapshot of assessment practice would not necessarily answer all of the questions raised above. Teachers devoted a great deal of attention to the difference between evaluating any particular assessment task and evaluating a teacher's entire assessment practice.

The participants next used the "Teaching" and the "Professional Development" chapters of the NRC Standards to glean additional criteria related to assessment practice. The teachers then incorporated the criteria into their original list of criteria; that list rapidly expanded from 23 criteria to 48. Participants then grouped the criteria into a meaningful set of categories that allowed them to reduce the number of criteria and remove redundancies. Participants struggled to identify the basic premises they were using to argue that the criteria represented good or "ideal" teacher assessment practice. The teachers identified six basic categories into which the criteria could be sorted:

- The purpose identified for the assessment
- Validity issues
- Assessments as learning tools
- Authenticity issues
- Equity issues
- Technical aspects of the assessments

The teachers raised several questions or issues about assessment practice. Those that resulted in the most discussion and debate among participants and staff were:

- Is it possible for an assessment task to address all categories?

- Is it necessary for an assessment task to address all categories?
- Are there good assessment tasks that address only some of the categories?
- How many categories does an assessment task have to address to be considered a good assessment?
- Are some categories more important than others?
- Do all assessments have to be authentic?
- Are the categories compatible?
- Does an assessment task have to address every item in a category to be considered a good assessment?

Table 2. Participants' Initial Criteria for Evaluating Assessment Tasks and Practice

1. Consistency; measures desired outcomes. If repeated, the same results would be obtained.
2. Equal emphasis on achievement and opportunity to learn science. Was the opportunity to learn available prior to measurement of achievement?
3. Authentic, real world application. Is there a match between what is being assessed and what is being taught? Does the assessment present science as scientists do science?
4. Assessment is ongoing, not simply one assessment at end of a chapter or unit.
5. Assessment is unbiased, fair to all students.
6. Does the teacher use a variety of assessment tasks resulting in a collection that is equally unfair to all?
7. Assessment produces sound inferences, objective not subjective.
8. Does the assessment task provide useful information to the shareholders?
9. Does the assessment encourage self-assessment or peer review?
10. Is the assessment developmentally appropriate?
11. Does the assessment provide for student involvement and interest?
12. Is there an opportunity to demonstrate knowledge that a scientifically literate adult would possess in order to make personal or societal decisions?
13. Is there an opportunity for a student to communicate understanding in a variety of performances?
14. Are the criteria for evaluation provided to the student along with the assessment task?
15. What evidence is shown that establishes a match between purpose and process?
16. How will the assessment task affect curricula and instruction?
17. Does the teacher make adjustments based on feedback from assessments?
18. How does the teacher demonstrate that he/she is a "reflective practitioner"?
19. Is the assessment tool also a learning tool?
20. Are there written plans that show the assessment is purposefully designed?
21. Are there explicitly stated purposes for the assessments?
22. Does the assessment indicate the extent and organization of students' scientific knowledge and the ability to communicate and use that knowledge?
23. Does the result accurately reflect both the opportunity to learn and student achievement?

For the final criteria-development activity, participants constructed an evaluation matrix that could be used to evaluate both the assessment tasks the teachers developed and their classroom assessment practices. The evaluation document is shown in Figure 2. In this way, teachers working from the NRC Standards (NRC, 1996) developed *their* own tool to evaluate *their* own assessment practice and tasks. Teachers used the evaluation criteria as a checklist to determine whether they had addressed or could address the various issues surrounding the assessment. Each of the teachers used this for self-evaluation of developed assessment tasks and included it with the documentation of each task. Teachers decided whether each task demonstrated the criteria exceptionally, specifically or not. For each evaluation category teachers were required to include a reflection that explained their thinking about the evaluation. This reflection process encouraged critical thinking about the quality of the assessment tasks and the identification of ways of enhancing the quality of the tasks.

OTHER PROFESSIONAL DEVELOPMENT ISSUES

Participants spent a significant amount of time differentiating the purpose of an assessment from the outcomes being assessed. Central issues included (a) determining who would use the information derived from the assessment and (b) what decisions the assessments would inform. Consumers of assessment information include teachers, students, administrators, parents, and members of the community. Assessment data informs a variety of decisions including the tracking of students, the determination of grades, the instructional evaluation and curricular choices, placement in special programs, and progression in grade sequence. These issues and the nature of the content and reasoning skills that exemplify science literacy determine the nature of the assessments teachers develop.

Quality control issues such as reliability, validity and generalizability were also raised and modeled for the participants. Participant knowledge about these issues was elicited by the assessment survey administered during the first workshop. An example of one teacher's perspective follows:

- Validity: a way of measuring the actual effectiveness of an assessment device or one part of that device.
- Reliability: the measure of the effectiveness of how well an assessment device or item truly meets its stated goal.
- Generalizability: how uniformly across a given population an assessment piece meets the stated goal.

Participants then analyzed their descriptions to construct a common understanding of these constructs. Illustrative vignettes were presented to the participants. During the resulting discussion, participants continued to refine their understandings of the terms and to determine how their assessment task development was affected by these concepts. They also explored the perspective on

Purpose	E	S	NS
Is the purpose of the assessment clearly articulated?			
Does the assessment provide useful information to the shareholders?			
Is the assessment a part of a repertoire of a wide range of performances?			
Validity			
Is there a match between the purpose of the assessment and the process used?			
Does the assessment measure valued outcomes?			
Key themes or concepts			
Cognitive skills			
Does the assessment provide an opportunity for the student to communicate what he/she knows?			
Does the assessment provide information for logical and sound inferences?			
Is the written plan for the assessment purposefully designed?			
Learning Tool			
Is the assessment an ongoing process?			
Is the assessment developmentally appropriate for the students with whom it will be used?			
Is there an opportunity for students to engage in			
Self assessment			
Peer assessment			
Authenticity			
Does the assessment reflect science as scientists do science?			
Does the assessment reflect the real world?			
Does the assessment engage students' interests?			
Does the assessment require the student to demonstrate knowledge to make personal and societal decisions?			
Is the assessment inquiry-based?			
Equity			
Is the assessment equitable?			
Is there evidence that the students have had the opportunity to learn the content and skills covered by the assessment?			
Technical Aspects			
Does the assessment task have a clearly written prompt?			
Are the criteria for scoring the assessment provided with the task?			
Are the scoring criteria appropriate for the task?			
Do the scoring criteria define a wide range of performances?			

Figure 2. Assessment Evaluation Matrix

the quality control issues of the *National Science Education Standards* (NRC, 1996) to clarify these issues.

Participants also addressed assessment bias by examining a number of assessment tasks for bias. An initial conceptualization of bias as the use of preferential language was replaced by the characterization of bias as decreased performance by any particular group for reasons unrelated to instructional activity. Participants were encouraged to examine student performance on the assessment tasks they had constructed to determine if any evidence of bias existed; however, little evidence of such activity was demonstrated in teachers' evaluations of their assessment tasks.

Because assessment tasks address reasoning processes as well as content knowledge, participants analyzed the Lactaid Task to determine the types of thinking skills they had to use to respond to the task. From this analysis, they generated a list of thinking and reasoning processes. This list was compared to others found in the science education literature to help participants better characterize the thinking and reasoning processes required by their own assessments. This characterization continued to be a difficulty for the teachers, and was the focus of several follow-up activities.

Few of the teachers expressed comfort with their abilities to construct adequate scoring guides at the first workshop. Holistic and analytic scoring guides were introduced early in the professional development program and a variety of rubrics were presented with the assessment tasks. These scoring rubrics were analyzed to determine the extent to which they addressed the reasoning skills and the content required for the assessment task.

In response to the teachers' concern about scoring rubric development, a follow-up activity required the teachers to construct their own responses to the "Plant in a Jar" task from the *National Science Education Standards* (NRC, 1996, p. 92). These responses were duplicated and added to a collection of responses collected from students at various levels. Each teacher developed a scoring guide and then practiced using the guide to evaluate the set of responses. The teachers then developed a scoring guide that represented the group consensus. Participants considered the developmental aspects of students' performances on various tasks to ensure that the scoring rubrics matched both the assessment task and the students who would be responding to the task.

The participants each presented their tasks and elicited a draft scoring rubric from their peers. This supported the teachers in their analysis of the requirements of the assessment tasks. The iterative nature of rubric development followed by task editing was promoted during several workshop sessions. Other workshop sessions allowed teachers time to share their task development and to reflect on changes in their perspectives. The teachers identified the need for such sharing early in the professional development program; teachers felt that they needed peer support as they modified both assessment tasks and practice. They identified several difficulties that were common to the group, including difficulties with integrating science content, inquiry, and reasoning skills into their assessments. Teachers found that they were frequently assessing science process and inquiry skills isolated from the content in the science discipline being taught. In addition, they reported difficulties in articulating the reasoning processes involved in the assessment tasks. This ongoing concern was addressed at follow-up workshops through further critiquing and analysis of assessment tasks. This indicates the need for long-term support and professional development in changing teachers' assessment practices.

CLOSING THOUGHTS

Linking assessment tasks to state and national documents and using assessment evaluation criteria based on the *National Science Education Standards* (NRC, 1996) makes assessment development "standards-based." In this project, teachers used the standards documents to describe how their own assessment tasks corresponded to characteristics of scientific literacy as stated in the various documents. For instance, descriptions of "science literacy" and "habits of mind" are clearly articulated in the *Benchmarks for Science Literacy* (AAAS, 1993). The *National Science Education Standards* (NRC, 1996) address assessment in more detail and develop the system and program aspects of science education. When incorporated into professional development programs, such documents can contribute significantly to teachers' development of standards-based assessment tasks.

ACKNOWLEDGEMENTS

Portions of this chapter are based upon work supported by a State of Indiana Commission for Higher Education (Grant No. 96-02) awarded to Shepardson, 1996. Any opinions, findings, and conclusions or recommendations expressed in this chapter are those of the authors and do not necessarily reflect the view of the State of Indiana Commission for Higher Education.

REFERENCES

American Association for the Advancement of Science. (1993). *Benchmarks for science literacy.* New York: Oxford University Press.

Bruner, J. (1988). Research currents: Life as narrative. *Language Arts,* 65, 574-88.

Carter, K. & Doyle, W (1989). Classroom research as a resource for the graduate preparation of teachers. In A.E. Woolfolk (Ed.), *Research perspectives on the graduate preparation of teachers.* Englewood Cliffs, NJ: Prentice Hall.

Frederick, P. (1990). The power of story. *AAHE Bulletin,* 43, 3.

Hardy, B. (1977). Towards a poetics of fiction. In M. Meek (Ed.), *The cool web.* London, England: The Bodley Head.

Kazemek, F. E. (1985). Stories of our lives: Interviews and oral histories for language development. *Journal of Reading,* 29, 211-18.

National Research Council (1996). *National science education standards.* Washington, DC: National Academy Press.

Schon, D.A. (1983). *The reflective practitioners.* New York: Basic Books.

EDITH S. GUMMER AND DANIEL P. SHEPARDSON

4. FACILITATING CHANGE IN CLASSROOM ASSESSMENT PRACTICE: ISSUES FOR PROFESSIONAL DEVELOPMENT

This chapter emphasizes two aspects of changing classroom assessment practice: the change process itself and impediments to change. The first portion of this chapter discusses the theoretical aspects of the difficulties teachers encounter as they reflect on and expand their assessment practice. The second portion of the chapter provides insight into the teacher impediments, internal and external, perceived or actual, for changing classroom assessment practice. These impediments are presented as issues to be considered in the professional development of science teachers' assessment practices.

ALTERNATIVE ASSESSMENT AND THE CHANGE PROCESS

Changing science teachers' assessment practice strikes at the heart of the difficulty of implementing change in an educational system. Changing assessment practice is no easy task for anyone involved in the functioning of schools. Assessment underlies all of the decisions that are made in schools, and all of the individuals involved in making those decisions use data from assessment instruments. Although it is not enough to examine only the changes that teachers must make to improve assessment, the teacher is central to the reform effort.

According to Fullan (1991), the key to the successful implementation of any change is the clear, coherent, and common meaning for all individuals involved about the purpose, the requirements, and the process of the change. Fullan (1991) stresses the involvement of the participants in the sense-making process:

> What we need is a more coherent picture that people who are involved in or affected by educational change can use to *make* sense of what they and others are doing (p. 4).

It is unclear whether science teachers have such a coherent, common meaning about what change in assessment is for, about what it involves, or about how it happens. Science teachers, for example, lack a common definition of assessment. Results from a survey completed by participants in one of our teacher enhancement projects demonstrates that teachers started the project with widely varying definitions of assessment, as revealed in the following statements:

Daniel P. Shepardson (ed.), Assessment in Science, 53—66.

- Assessment is a tool to measure a student's understanding of material taught.
- Assessment is an attempt to determine what students know and what they can do.
- Assessment is any means of measuring a student's class work, knowledge, understanding and/or ability to use that learned information. Assessment can also be use to evaluate many other aspects of the class including but not limited to curriculum, teacher-project effectiveness.

Terms such as "alternative" or "authentic" assessment had even fewer common definitions in our project. For instance, "alternative assessment" was characterized as follows:

- An alternative assessment is a method of evaluation of student performance other than a standard question/answer method.
- Getting away from multiple choice based tests or lectures, varying activities, methods of teaching, maximizing credibility/validity of assessments used in teaching.
- Any assessment device other than standard forms and formats (quiz, test, etc.) or questions within the quiz or test that has a non-standard format, usually related to higher-level learning.
- A non-traditional way to determine student's understanding of a topic.
- A task that provides information/data as to what each student knows and is able to do.

This variability in understanding alternative assessment highlights the lack of any shared understanding about what the change process means to science teachers who are at the center of such change. Teachers responded in an even more variable manner when asked to define what "authentic assessment" meant to them:

- Authentic assessment is an assessment process that accurately measures the concepts or skills the assessment was designed for.
- Is the assessment task valid and reliable?
- An assessment that measures standards and objectives.
- An assessment that requires a real world approach for resolution.
- Real world problems/scenarios that are used as prompts to demonstrate knowledge.

These differences in characterizing the concepts that underlie the process of assessment are not trivial. They demonstrate that those who are being required to implement the changes in assessment practice do not start the process with the same knowledge base about assessment in science. Teachers lack a common, coherent picture of what assessment means in the context of their own practice. Change, then, will mean different things to each individual.

The meaning of an educational practice and of any change in that practice is only the first of several factors Fullan (1991) identified as being central to change. Marris (1975) adds to this picture by arguing that all real change involves loss, anxiety, and struggle. Identifying what teachers will lose as they change their assessment practice is crucial to enhancing assessment practice. The anxiety of giving up a practice with which a teacher is familiar must be addressed, as the teacher is encouraged to change. In turn, the areas around which a teacher will struggle as he or she is engaged in change must be identified and addressed.

In order to understand these factors, facilitators must also understand both the objective and subjective meanings of educational change (Fullan, 1991). The objective meanings are multidimensional and involve three components or dimensions: new or revised materials, new teaching approaches, and new beliefs about assessment. Teachers may be presented with changes in assessment along each of these dimensions. For instance, new assessment tasks may accompany new textbooks and curricular materials. Teachers may be required to incorporate new forms of assessment by school, district, or higher administrative authorities. Finally, teachers may come to believe that newer forms of assessment better serve both their teaching and their students' learning. Although each of these dimensions involves change in practice at some level, it is through a change in beliefs that teachers incorporate change at a core level.

This third dimension of change serves as a bridge to what Fullan (1991) calls the subjective meaning of change, the teacher's determination of what change in assessment practice means for him or her. A central aspect of the subjective meaning of change involves what Hubermann (1983) characterizes as the concept of "classroom press." This refers to immediacy and concreteness, multidimensionality and simultaneity, adapting to ever-changing conditions or unpredictability, and personal involvement with students. According to Hubermann (1983) and Crandall & Associates (1982), this press significantly affects what teachers focus upon and teacher energy for engaging in real change. Doyle and Ponder (1977, p. 78) argue that most teachers follow a "practicality ethic." This means that a teacher seeks to ascertain how a change in assessment practice will affect students and how well it will match with their current assessment approach. Another aspect involves the clarity of the procedures through which the change would be implemented. Another aspect involves the teacher's estimation of the cost of change in terms of time, energy, and investment of self-esteem in relation to future benefits. On this view, teachers evaluate new assessment approaches based on the time and energy involved in developing, implementing, and using such assessments, as well as the degree to which the new assessment approach contradicts current practice.

Changing assessment practice is not a superficial fix; it is not the simple administration of new assessment tasks or the acquisition of new pre-packaged assessments. Instead, change involves reflecting on and adapting many aspects of classroom practice including curricular choices, instructional techniques, and classroom management. In other words changing science teachers' assessment practice goes beyond simply changing classroom assessments, to understanding the relationship between assessment, curriculum, pedagogy, and the learner. As a result, teachers run the risk of engaging in two forms of non-change (Fullan, 1991), "false clarity" and "painful unclarity." False clarity means thinking that one has changed a practice when one clearly has not. Teachers who demonstrate false clarity would use assessment tasks that may appear alternative in form, but that fail to expand ways of gathering evidence about what students know and can do in science. Teachers can also experience false clarity by administering alternative assessments without using the results for any of the various purposes such results should inform. This is change in the superficial aspects of assessment practice.

Painful unclarity involves forcing teachers to implement changes they do not really understand. This occurs when the issues surrounding the use of alternative assessments are not clearly and fully developed with the teachers. This highlights the notion that change in assessment practice is not trivial. It involves understanding the important goals of the study of the science discipline, modification of instructional practices to support those goals, understanding that assessments provide information for a variety of purposes, and the use of evaluation techniques that are novel to many science teachers. The important goals of science have been well characterized in terms of the content and skills that students should know and be able to do (e.g. *Benchmarks for Science Literacy*, AAAS, 1993; *National Science Education Standards*, NRC, 1996). Curricular changes and matching instructional modifications, however, have been slower to appear. In addition, our understanding of the cognitive components required by these essential skills lags behind our ability to reliably measure such skills in science.

Fullan (1991) identifies six implications of the subjective and objective realities of educational change. These include:

1. The soundness of proposed changes.
2. An understanding the failure of well-intentioned change.
3. Guidelines for understanding the nature and feasibility of particular changes.
4. The realities of the status quo.
5. The deepness of change.
6. The question of valuing.

Each of these implications must be addressed in evaluating how change in assessment practice can be brought about. First, teachers need to see that changing assessment practice is worthwhile in terms of both the effort that change requires and the results in student learning and motivation. They need to be convinced that changes in assessment practice are more than a simple fad that will pass in time. They also need to see that changes in assessment practice at one level of the educational system are reflected at other levels. For instance, high school teachers will be reluctant to change from assessments primarily involving multiple-choice tests if their students are evaluated solely with such instruments in college science courses. Another concern of teachers is the time that such expanded assessment tasks demand. Teachers are concerned that they will not provide their students with exposure to science content that they will be expected to know in subsequent science courses.

Understanding how change in assessment practice is brought about involves understanding efforts that have not succeeded. Efforts in other countries such as England, and other states, such as California, provide awareness of the complexity of change, and highlight pitfalls that teachers and schools might avoid. For example, Nuttall (1992) reported that England's Assessment of Performance Unit (APU) resulted in teachers learning more about children's achievements; however, the performance tasks are more demanding of teachers' time and energies. British teachers also raised concerns about classroom management, and some viewed the performance assessments as less rigorous.

Sustained interest in assessment is a recent phenomenon. Assessment vies for attention with other aspects of teacher preparation or enhancement in science education. Numerous studies have also examined the feasibility of changing assessment practice at the classroom level. Information from these studies needs to be incorporated into professional development programs for both inservice and preservice teachers. Identification of the ways in which assessment permeates curricular and instructional decisions is needed to clarify the means of changing practice.

The study of the current assessment practices of science teachers has largely focused on grading methods though there are some notable exceptions (Stiggins & Conklin, 1992; Kane, Khattri, Reeve & Adamson, 1997). Studies that examine science teachers' assessment knowledge and beliefs and classroom practice are slowly emerging as a needed focus of further research. Findings show that teachers need to have the opportunity to observe other teachers as they incorporate alternative assessments into the daily activities of their classroom. Teachers also need to be aware of the complexity of changing assessment practice. One teacher stated the following in the middle of a professional development program during which she designed alternative assessment tasks:

> If I'm going to assess my students' understanding like this, I'm going to have to change my instructional strategies. Developing alternative assessment tasks requires developing alternative instructional strategies.

Recognition of the significant effects of change on other educational practices should not come as a surprise to teachers. Finally, science teachers need the opportunity to explore the value of changing to alternative assessment practices. Teachers need the time and support to reflect on and evaluate those who will benefit from the change and the effect of such change on the various outcomes of teaching. Fullan asserts that:

> In theory, the purpose of educational change presumably is to help schools accomplish their goals more effectively by replacing some structure, programs and/or practices with better ones (1991, p.60).

He identifies two types of goals. First, students must be educated in various academic or cognitive skills and knowledge; second students must acquire the individual and social skills and knowledge necessary to function occupationally and socio-politically in society. Teachers need to explore the significance of these goals for their own teaching and identify how or if alternative assessments support their attainment.

Finally, Fullan (1991) identifies two more aspects of change that are crucial to success: the phenomenology of change and the cognitive load of change. By phenomenology of change, he means determining how people actually experience change as opposed to simply articulating how change is expected to occur. Neglect of this actual experience is central to the failure of change. The cognitive load of change involves understanding the various aspects that are affected by the change. Fullan concludes:

> Thus, on the one hand, we need to keep in mind the values and goals and the consequences associated with specific educational changes; and on the other hand, we need to comprehend the dynamics of educational change as a sociopolitical process involving all kinds of individual, classroom, school, local, regional, and national factors at work in interactive ways. The problem of meaning is one of how those involved in change can come to understand what it is that should change and how it can be best accomplished, while realizing that the what and how constantly interact and reshape each other (1991, p.5).

These two aspects of change are important whether the source of change is internal, involving the individual teacher, or external and directed toward a whole school, district, or state.

Educational change is no longer viewed as a linear process (Stiegelbauer, 1996). Rather it involves overlapping waves of initiation, implementation and institutionalization. Stiegelbauer (1996) argues that teachers need information and leadership in order to change their assessment practices. These needs can be supported by opportunities for professional development coupled with policies that support change. One without the other leads to the non-change in assessment practice. Stiegelbauer (1996) further identifies dual purposes for assessment in bringing about change. On the one hand, alternative assessments are the focus of a specific innovation in practice in the classroom and beyond. On the other hand, results from such assessments provide evidence that change has occurred in other aspects of educational practice.

It is important to realize that change in assessment practice represents what Cuban (1988) has characterized as second order changes rather than first order changes. First order changes involve more superficial aspects of classroom practice and school systems. Second order changes involve change in the organization of schools and the modification of how people work together to attain the goals of education. Changing assessment practice may integrally involve teachers, but it also requires that the beliefs, attitudes, and practices of students, parents and school administration change. These second order changes are deeper, and they are harder and slower to initiate and implement, requiring sustained attention in order to be realized.

ISSUES TO CONSIDER IN PROFESSIONAL DEVELOPMENT FOR THE PURPOSE OF CHANGING ASSESSMENT PRACTICE

This portion of the chapter provides insight into specific impediments or issues teachers may face in changing their classroom assessment practices. These issues emerged from our collaboration with teachers and are summarized in Table 1. Each should be considered when designing professional development programs to change assessment practice.

Table 1. Teacher Constraints to Developing and Implementing Alternative Assessment in Science

---

1. Developing, implementing, and using alternative assessments is time intensive.
2. Instructional strategies must be considered in light of alternative assessments if change in assessment practice is to occur.
3. All stakeholders, such as parents, colleagues, and administrators, need to be apprised of the impact of alternative assessment on student leaning.
4. Students need to become familiar with alternative assessments.
5. What a teacher considers as assessment of students' "knowing the material" must change from factual recall to multiple student traits through multiple sources.
6. Long-term support outlasts short-term efforts for professional development.
7. Teacher collaboration supports change in assessment practice.
8. Classroom context for assessment task development matters.
9. School leadership and support may conflict with classroom assessment priorities.
10. Knowledge about assessment task development and practice.
11. Change requires access to information and resources.

Issue 1: Alternative Assessment is Time Intensive

Perhaps the most common constraint to the development and implementation of classroom-based alternative assessment tasks was time, or more accurately, the insufficient amount of time to develop, implement, and use assessment tasks. Because of time constraints, teachers often used assessments provided by textbook publishers; these emphasized low-level thinking and factual recall (American Association for the Advancement of Science [AAAS], 1998). Time concerns reported by teachers involved in our assessment projects included: 1) time to develop and revise assessments, 2) time taken from classroom instruction to implement assessment, and 3) time required to score assessments. These concerns are significant and have no simple remedy. Although one solution is to make assessment seamless with instruction, this will not completely eliminate the increased time needed to implement alternative assessments in the classroom. In the end it comes down to a fundamental change in a teacher's understanding and beliefs about what it means to learn and to assess.

We found that teachers' development of assessment tasks, as well as the quality of their assessment tasks was greatly hindered as the amount of time available to develop assessment tasks declines. When teachers received sufficient time to develop classroom assessments, to collaborate with colleagues, to dialogue with peers about assessment practice, to research and reflect on classroom assessments, and to integrate instruction and assessment, they made substantial gains in their assessment knowledge. This resulted in substantial change in assessment practice. Teachers also needed time to share experiences, knowledge, and beliefs about assessment. Providing teachers the opportunity to engage in dialogue with colleagues and visit classrooms implementing alternative assessments is essential to fostering change in practice and establishing a community of teacher-learners. In sum, professional development must include opportunities for teachers to visit teachers.

Issue 2: Instructional Strategies Must be Considered in Light of Alternative Assessments if Change in Assessment Practice is to Occur

Any effort to shift teachers' practice toward the use of alternative assessment tasks in science requires that teaching strategies change to align with assessment. Assessment must be viewed in the context of the curriculum and instruction. If, for example, an assessment task requires students to demonstrate the ability to assemble an electric circuit, then instruction must provide students the opportunity to learn how to construct an electric circuit. Further, if understanding science content and the ability to deal with new problems in creative ways are goals of teaching, then the loss of content coverage is a non-issue and alternative assessment is appropriate. If, however, the goal of teaching is breadth of content and how to solve problems, then traditional assessment is appropriate. Teachers and schools desiring to use alternative assessment must determine their goals for student leaning before adopting alternative assessment strategies; otherwise, the outcomes will not be favorable.

Issue 3: All Stakeholders, Parents, Colleagues, and Administrators, Need to be Apprised of the Impact of Alternative Assessment on Student Learning

Many teachers in our project perceived conflicts with parents or administrators as a central concern to the implementation of alternative assessment in their classrooms. Parental issues ranged from fairness in grading to preparation for test-taking at the next grade level; this was often a concern of parents with children in high school. Student conflicts, on the other hand, were not perceived as constraints if students had practiced completing alternative assessment formats, had gained experience with such formats, and had become aware of the expectations for completing such assessments. This may be due to teacher perception that alternative assessment was more beneficial to learners.

Professional conflicts identified by teachers tended to center on concerns about other teachers, about school administration, and about curricular goals. Teachers were also concerned that other teachers did not value alternative assessment and viewed their use of such assessments negatively because they were different. Conflicts with administrators tended to center around differences from other teachers and the use of assessments that were perceived to be subjective. Curricular concerns related primarily to time; alternative assessments required more time to implement and left less time for covering the school district curriculum. This results in a conflict between classroom practice and school policy.

Central to avoiding these dilemmas is clear communication to all stakeholders about assessment purposes, expectations, and quality. If parents and administrators perceive assessments to be productive and meaningful then confrontational situations may be avoided. Communication should also emphasize the clear definition of what is being assessed and how, for example, the scoring rubric needs to be understandable and its use valid and reliable. Examples of past student work need to be kept and shared with all stakeholders to communicate the purpose, expectations, and quality of the assessments during parent-teacher conferences or back-to-school night. Teachers might also be encouraged to send letters to parents at the beginning of the school year to inform them about the use of alternative assessments.

Issue 4: Students Need to Become Familiar with Alternative Assessment

Students may pose an initial difficulty when alternative assessment tasks are first implemented. Whether this should be attributed to different learning styles, to unfamiliarity with the new assessment strategy, or to the need of new ways of thinking is unclear. In the most extreme case, students may become barriers to the implementation of alternative assessments because such assessments are different and more labor-intensive (AAAS, 1998). Regardless of the cause, assisting students in learning how such assessments are to be completed may promote the use of alternative assessment. This means that alternative assessment needs to be phased in over time, because students must practice the new and different assessment formats. Students need to become familiar with the new expectations and relearn what it means to be assessed. Efforts to adopt alternative assessment without allowing time for student adjustment will progress with difficulty. In addition, instruction must align with assessment so that students gain experience and practice in completing the alternative assessments. The teacher should discuss with students how the assessments will be done and should share examples of student work from prior years. Students should also be provided opportunities to contribute to the design, implementation, and scoring of alternative assessments (AAAS, 1998).

Issue 5: What a Teacher Considers as Assessment of Students "Knowing the Material" Must Change from Factual Recall to the Assessment of Multiple Student Traits Through Multiple Sources

In order to change assessment practice, teachers have to shift from a content focus to a focus that stresses understanding and the pathways to that understanding; that is, teachers must consider multiple learner traits. This shift requires the development of assessment tasks and scoring systems that are sensitive to both teaching goals and to multiple student traits. The process is tedious at best and requires time to refine the assessment tasks and the scoring systems appropriate for assessing the desired traits.

How does one facilitate change in views toward the development of assessment tasks? We have found that teachers who critically reflect on their practice achieve this desired outcome. Critical reflection based on real classroom practice and the activity of learners may result in cognitive dissonance about practice by the teacher. This may lead to change in practice (Huberman, 1995; Nelson & Hammerman, 1994). For example, when teachers share their experiences with developing and using scoring rubrics, they may discover they are inadequate for assessing student reasoning abilities or with assigning scores. Teacher reflection on the assessment tasks developed promotes critical examination of methods of assessing and teaching. This is necessary for teachers to if the merit of a given scoring rubric is to be determined in light of student work and the classroom situation. When teachers realize that students have not necessarily learned what was taught, they tend to reconsider their own fundamental ideas about teaching and learning (Darling-Hammond, Ancess, & Falk, 1995). Only through this reflection process can assessment tasks and scoring systems be revised in a way that will strengthen the assessment of students.

Issue 6: Long-term Support Outlasts Short-term Efforts

Without long-term support many teachers will abandon their efforts to develop alternative assessment tasks, and will continue with their current practice. We have found that many teachers do not develop new assessment tasks outside the scope or boundaries of the professional development program. Short-term support will result in limited change; however, few institutions commit to long-term support because of the expense. Most state and federal agencies fail to support long-term professional development activities that deliver intensive support.

Issue 7: Teacher Collaboration Supports Change in Assessment Practice

Perhaps the most significant factors in supporting change in teacher assessment practice have been teacher collaboration and peer support. Teacher collaboration involves teachers working together to develop similar assessment tasks, while peer support involves teachers critiquing and discussing assessment tasks or practice. Teachers may not, however, be as critical as they should be or as we would like.

To promote teacher collaboration, professional development programs could require the formation of teacher teams to develop assessment tasks. Peer support allows teachers to work in small groups to share and critique assessment tasks and student work examples, and to discuss ways of improving the assessment task. We found it helpful to first brainstorm criteria for critiquing or evaluating assessment tasks. This process starts in small groups and expands to whole group discussion and consensus building about the criteria to be used. The criteria used for critiquing each teacher's assessment task then emerge from the teachers themselves; they become the teachers' criteria and not those of the professional development staff.

Issue 8: Classroom Context Matters

It goes without saying that any assessment task must be developed and understood in the context of the teacher's classroom and school environment. Teachers need to develop, field test and reflect on their assessment tasks in the context of their own classrooms, and they need to revise the assessment tasks based on how well the tasks fit with their classrooms. The implication for professional development is straightforward; professional development must incorporate the teacher's classroom, the teacher's world and real students.

Issue 9: School Leadership and Support May Conflict with Classroom Assessment Priorities

In schools and districts where administrators support change in assessment practice, teacher change is more likely; however, most school and district administrators support individual teachers who are interested in participating in professional development activities to improve assessment practice. Although this may improve the assessment practice in one classroom, it fails to bring about school district-wide change in practice. In some situations, the support of individual teachers leads to perceived or actual conflicts between teachers in schools. It also sends confusing messages to students about what is valued in science and what the expectations are.

Without firm administrative support and leadership, change in classroom assessment practice will bend to accommodate the changing policies and priorities of the school or school district. We recall one situation in which a high school teacher in one of our assessment development programs removed herself from participating. She did so because the school district's priority changed to emphasize the alignment of science content with the new state science standards or competencies. As department chair she was required to evaluate the match between the items on each teacher's classroom test with the state high school competencies. This illustrates a significant mismatch between the development of assessments that expand measures of science learning and the accountability aspect of assessment.

Issue 10: Knowledge about Assessment Task Development and Practice

In our experience, teachers had the most difficulty in developing assessment tasks that assessed students' scientific reasoning and thinking abilities. Teachers were fairly comfortable in designing assessment tasks that focused on science process skills and procedural knowledge. Teachers were somewhat more confident in designing assessment tasks that measured students' conceptual and factual understandings. Developing assessment tasks that incorporated and integrated reasoning processes with conceptual and factual understanding, however, was extremely difficult. Further, teachers encountered difficulties in developing scoring systems that captured students' scientific reasoning and thinking abilities, and that differentiated student abilities.

Many teachers felt they lacked the technical knowledge to design these kinds of assessment tasks, to develop scoring systems and to understand different assessment formats. These knowledge-based issues have direct implications for what Kane, Khattri, Reeve, and Adamson (1997) have referred to as the pedagogical usefulness and technical rigor of assessments. Pedagogical usefulness refers to how well an assessment is integrated with instruction, while technical rigor links an assessment with science content and performance standards.

Reliability and validity issues for teachers focused on the classroom use of assessments. Reliability, for teachers, was essentially an issue of fairness and consistency in scoring assessments. Issues of validity tended to emphasize the appropriateness of content and processes; that is, did the assessment reflect what was taught in terms of the science content and processes of science? These views of reliability and validity were limited in that they did not address issues of equity, quality, administrative procedures, or generalizablity; such aspects involve inferences made about student ability based on student performance on a given assessment task.

Issue 11: Change Requires Access to Information and Resources

One of the constraints to changing teachers' classroom assessment practice in science is access to information and resources to develop a knowledge base about assessment. Without resource materials to build from, change in teacher assessment practice is constrained. Teachers gain access to information and resources through participating in professional development programs that promote an intellectual base for assessment, and that provide access to national reform documents (e.g., *National Science Education Standards*, NRC, 1996; *Benchmarks for Science Literacy*, AAAS, 1993), books, articles, and websites about assessment. These resource materials may assist teachers in constructing frameworks for their assessment practice, in identifying levels of proficiency, and in providing insight into developing assessment tasks. Information may also be obtained through interactions with experts in assessment development and through interactions and site visits with colleagues from other schools where alternative assessment tasks have been developed and implemented. Access to colleagues can provide a support base for discussing assessment practice and can reduce the sense of isolation.

CONCLUDING THOUGHTS ON THE PROFESSIONAL DEVELOPMENT OF SCIENCE TEACHERS' ASSESSMENT PRACTICE

The goal of professional development programs in science assessment is to develop teachers who are assessment-literate; this results in teachers who are knowledgeable about assessment task development and assessment practice, and who engage in changing their assessment practices. Stiggins (1991) suggests that assessment-literate teachers are knowledgeable about the following:

- Four achievement categories: 1) substantive subject-matter knowledge, 2) thinking skills to be demonstrated, 3) desired behaviors to be exhibited, and 4) products with specific attributes.
- Assessment decisions made; how those decisions relate to effective teaching and learning, and what kind of data best informs decision-making.
- Attributes of sound data, factors that reduce the quality of data, and strategies for avoiding such pitfalls.
- Three types of assessments: paper-and-pencil instruments to assess knowledge and thinking skills, assessment based on observation and judgment that focuses on important behaviors and products, and interaction with students to obtain direct insight into students understandings and learning.
- Strengths and weakness of each of type of assessment.

Science teachers who are assessment-literate develop and use assessments that communicate what is valued clearly and specifically. Such teachers use assessments that reflect precisely defined performance targets; they realize the importance of fully sampling performance, they are aware of extraneous factors that can interfere with assessment results, and they understand when assessment results are not meaningful (Stiggins, 1991).

REFERENCES

American Association for the Advancement of Science. (1993). *Benchmarks for science literacy.* New York: Oxford University Press.

American Association for the Advancement of Science. (1998). *Blueprints for reform: Science, mathematics, and technology education.* New York: Oxford University Press.

Crandall, D. & Associates. (1982). *People, policies and practice: Examining the chain of school improvement* (Vols. 1-10). Andover, MA: The Network, Inc.

Cuban, L. (1988). A fundamental puzzle of school reform. *Phi Delta Kappan, 70*(5), 341-344.

Darling-Hammond, L., Ancess, J., & Falk, B. (1995). *Authentic assessment in action: Studies of schools and students at work.* New York: Teachers College Press.

Doyle, W., & Ponder, G. (1977-78). The practicality ethic in teacher decision making. *Interchange, 8*(3), 1-12.

Fullan, M.G. with S. Stiegelbauer. (1991). *The new meaning of educational change.* New York: Teachers College Press.

Hubermann, M. (1983). Recipes for busy kitchens. *Knowledge: Creation, Diffusion, Utilization, 4,* 478-510.

Huberman, M. (1995). Networks that alter teaching: Conceptualizations, exchanges and experiments. *Teachers and Teaching: Theory and Practice, 1*(2), 193-211.

Kane, M.B., Khattri, N., Reeve, A.L., & Adamson, R.J. (1997). Assessment of student performance: Studies of education reform. Washington, DC: Office of Educational Research and Improvement, U.S. Department of Education.

Marris, P. (1975). *Loss and change.* New York: Anchor/Doubleday.

National Research Council [NRC]. (1996). *National science education standards.* Washington, D.C.: National Academy Press.

Nelson, B.S. & Hammerman, J.K. (1994). *Reconceptualizing teaching: Moving toward the creation of intellectual communities of students, teachers and teacher educators.* Newton, MA: Center for the Development of Teaching, Education Development Center.

Nuttall, D.L. (1992). Performance assessment: The message from England. *Educational Leadership*, *49*(8), 54-57.

Stiegelbauer, S. (1996). Change has changed: Implications for implementation of assessments from the organizational change literature. In M. B. Kane & R. Mitchell (Eds.), *Implementing performance assessment: Promises, problems, and challenges* (pp.139-160). Mahwah, NJ: Lawrence Erlbaum.

Stiggins, R.J. (1991). Assessment literacy. *Phi Delta Kappan, 72*(7), 534-539.

Stiggins, R. J. & Conklin, N.F. (1992). *In teachers' hands: Investigating the practices of classroom assessment.* Albany, NY: State University of New York Press.

DANIEL P. SHEPARDSON

5. THINKING ABOUT ASSESSMENT: AN EXAMPLE FROM AN ELEMENTARY CLASSROOM

As teachers begin to think differently about teaching and children's science learning, they must also begin to think differently about assessing learning. The assessments teachers develop and use greatly impact science learning (American Association for the Advancement of Science [AAAS], 1998). This chapter shares the thinking processes that a fifth grade teacher, Laura, and I engaged in as we talked about assessment issues and practice. This process was facilitated by shared reflection on classroom assessment practice and collaborative dialogue through which we constructed a consensual understanding about science assessment. The collaborative process facilitated Laura to think differently about assessment and provided her with the impetus to change her classroom assessment practice.

The collaboration with Laura was a result of her participation in one of my professional development programs. From our interactions and shared experiences in the program, Laura became interested in assessment and in changing her assessment practice. We mutually agreed to look at and talk about her assessment practice in an upcoming unit on electric circuits. As part of our collaboration, Laura allowed me to audiotape our conversations, observe her classroom assessment practice, take field notes, and use examples of her classroom-based assessments, which are presented in this chapter. Many of our conversations took place in Laura's classroom at the end of the school day prior to, during, and at the end of the electric circuit activities. At the end of each meeting we often identified topics or planned assessment tasks to discuss at our next meeting.

Both the chronology and the assessment examples illustrate the highlights of our collaborative thinking processes about assessment in this particular instructional context. Issues about assessment and assessment practice are raised throughout the chapter; although some of these issues were encountered throughout our dialogue, they are discussed here only once to reduce repetition. The structure of this chapter reflects the issues and content of our conversations and do not necessarily follow the historical order of our conversations. Several of the tools for thinking about assessment practice that emerged from our collaborative dialogue are presented more formal. The chapter concludes with a presentation of several issues about science assessment that should be incorporated into any dialogue about classroom assessment.

Daniel P. Shepardson (ed.), Assessment in Science, 67—82.

REFLECTING ON ASSESSMENT PRACTICE

Starting With the Known

Laura and I began to think about science assessment by reflecting on her current assessment practice; we identified the instructional goals and purposes of her assessments (i.e., the intended assessment), we looked at the types and formats her assessments used (i.e., the implemented assessment), and we characterized the nature or domain of student learning actually assessed (i.e., the experienced assessment). By coming to a shared understanding about current assessment practice, the need for change was linked to the context of the classroom and provided a foundation for building a new understanding of practice.

Laura used two assessment pieces in the electric circuit unit that involved children in about three weeks of instruction; the first week emphasized simple circuits, the second week series circuits, and the third week parallel circuits. Before talking about the assessments, it is necessary to briefly describe the instructional setting as a context for thinking about the assessment. The electric circuit unit involved children in inquiry activities, in making predictions, in testing their ideas and in exploring batteries and bulbs. Student activities ranged from drawing how they would light a bulb and then testing these predictions to depicting a battery-to-bulb connection and then manipulating the materials to test their ideas. Students also constructed series and parallel circuits. The unit ended by having students construct their own circuits through trial-and-error, using multiple batteries and bulbs and various bells and switches; students then drew a blueprint for other student groups to follow as they constructed the "new" circuits. Laura's assessments were administered at the end of the unit as summative assessments. One assessment required children to predict whether the bulb(s) shown in different circuit diagrams would light and to draw the wire connections between a battery and bulb(s) such that the bulb(s) would light, and the other assessment was a traditional textbook-based quiz (Table 1).

When asked to explain her instructional goals and purposes for assessment, Laura stated she wanted children to learn about the different types of electric circuits (i.e., simple, series, and parallel) and how circuits worked, how electricity flowed in and out of the battery and through the light bulb, bell or other electrical devices in the circuit. Her purpose for assessment was to determine what children had learned about electric circuits from their classroom experiences; this purpose was based on her need to grade the students. In describing her first assessment, where children drew the wire connections, Laura said that it informed her about children's understandings of electric circuits and how electricity flowed in a circuit:

> How they draw the wire to connect the battery and bulb tells me that they know how to connect a bulb and that [they] know how the electricity moves in and out of the battery, in a circuit, a path.

Table 1. Laura's Electric Circuit Quiz

A. Match the word with the definition by writing the letter of the word next to the definition.

1. A circuit in which electricity can flow in only one path.	a. Electric current
	b. Battery
2. An object that electrons can easily move through.	c. Circuit
3. Changes chemical energy into electrical energy.	d. Parallel circuit
4. A path that electrons can move through.	e. Series circuit
5. A circuit that allows electricity to flow in more than one path.	f. Conductor
	g. Static electricity
6. The movement of electrons.	

B. Describe a closed circuit.

C. Explain how a series circuit is different from a parallel circuit.

The second assessment (Table 1) ". . . tells me what students know about series and parallel circuits . . . about other aspects of circuits . . . like if they are closed or open." Thus, these assessments were designed to evaluate students' abilities to make an electric circuit and to determine if students understood the different types of electric circuits.

Our next step was to critically think about Laura's current assessments. Some of the questions we talked about were:

- Do they assess everything we value and want students to know and do in science?
- Do they really assess what we think they assess?
- What do they tell us about what students have learned? What students know and can do in science.
- Are there ways we can improve these assessments?

To accomplish this we constructed an assessment analysis matrix similar to the one shown in Table 2. As we began to talk about Laura's current assessment practices, it became clear that they were not assessing children's process skills and procedural understandings; domains valued by Laura, but not made explicit in her assessments. Instead, her assessments emphasized low-level thinking and factual knowledge or recall, and not conceptual understanding. Laura's thoughts:

> Well I guess they're maybe not really assessing science process skills. I sort of assumed that by [having the students] draw I was getting at the science process skills . . . their [students'] ability . . . understandings about how to build a circuit. . . . Well, again [in reference to the quiz] I was trying to determine what students understood about electric circuits, but perhaps I need to look at those questions again . . . I took them pretty much from the textbook . . . but maybe they are low-level. Maybe I could use Blooms levels to improve the questions?

Table 2. Assessment Analysis Matrix

Questions	Assessments		
	Drawing Wires	**Quiz**	
Does the assessment align with what we value and with what students do?	*Not completely. Does not require students to actually construct a circuit with materials.*	*Not completely. Does not require students to actually construct series or parallel circuits with materials.*	
Does the assessment assess what we want to assess about student learning?	*Not completely. Does not assess students' understanding of electric current. Does provide some indication of students' understanding of connections.*	*Not completely. Assesses only factual understanding and not conceptual understanding.*	
What does the assessment tell us about what students learned? What they know and can do.	*Provides an indication about what students know about simple circuits.*	*Provides an indication about what factual knowledge students learned. What vocabulary they have mastered.*	
Are there ways we can improve the assessment?	*Yes. Require students to explain in writing their reason for why they drew the wires the way they did. Have students use materials and make a circuit, explaining the reason behind the way they made the circuit.*	*Yes. Develop higher-level thinking questions based on Bloom's taxonomy.*	

Note: One column is intentionally left blank.

Laura had also assumed that if students were able to draw wires to connect a battery and bulb, they also understood electric current, which, according to the science education literature, is far from the case. From our conversations, we began to think about ways of improving these existing assessments: we might require students to write explanations for their wire connections, have students actually use a battery, bulb, and wire to make a circuit, and explain the reason behind the way they made the circuit, we might develop higher-level thinking questions based on Bloom's (1956) taxonomy.

After we had thought and talked about ways of improving existing assessment activities, I asked Laura to think about when these assessments were delivered or completed by students. After a puzzled look she replied, with further puzzlement, ".

. . at the end, after they have completed the activities." Our talk continued, focusing on when students should be assessed. We questioned whether students should be formally assessed only at the end of an instructional unit or whether they should be assessed throughout the unit (i.e., formative assessment). Should there be a pre-assessment or diagnostic assessment? These focused our talk to address ways in which assessment practice could be enhanced based on the delivery of assessment tasks. This naturally flowed into talking about evaluating students or "grading" as Laura called it (see evaluation and grading section below).

Changing Assessment Practice

Following on our earlier questions and conversations, our talk shifted to ways of viewing student assessment as diagnostic, formative, and summative. Through diagnostic assessment the instructional unit can build on children's ideas, engage children in thinking about the content, and provide a lens for looking at changes in children's understanding and thinking over time. Formative assessment provides evidence of children's progress throughout the lesson, and in a cumulative manner, provides evidence of children's science learning. Formative assessment can also inform and direct instruction, guiding teachers' pedagogical moves. Summative assessment provides a picture of what students know and can do at the end of the unit. Laura and I also talked about the assessment task formats one might use to assess children's understandings about electric circuits: should all the assessment tasks be paper and pencil? Should the task be individual or group-based? Should multiple-choice questions be used? Should children's products alone be assessed? By thinking about assessment in these ways, a more holistic view of assessment may be constructed; assessment begins to be viewed as a system within the classroom instead of a series of individual events or happenings. The National Research Council's [NRC], Teaching Standard C states that ". . . assessment tasks are not afterthoughts to instructional planning but are built into the design of teaching" (NRC, 1996, p. 38). Viewed as a system, assessment permits the construction of a better picture of students' science learning; it profiles student performance and abilities both within an unit of instruction and over time (see section on profiling below).

In addition, we talked about the research on children's ideas about electric circuits (Table 3) and how this might inform us not only about new assessments, but also about instructional activities. The research literature on children's ideas provided us with a starting point for thinking about assessment tasks and how to look at and score children's performances. For example, the research indicates that children tend to construct one of four models of electric current (Osborne, 1983) and that the "clashing currents" or nonrecursive model is the most predominantly held by elementary age children (Osborne, 1983 & Solomon, 1985). This means that assessments must elucidate these electric current models. By developing assessments that identify the electric current model(s) held by children, pedagogy may be informed and change in children's developing understandings can better be determined.

Table 3. Summary of Research Findings on Children's Ideas about Electric Circuits

Author	Findings
Osborne (1983)	Children construct one of four models of electric current: nonrecursive, clashing currents, unidirectional current consumption, and scientific.
Shipstone (1985)	Children use multiple models of electric current to explain different circuits; children hold several models of electric current.
Shepardson and Moje (1994)	Children's understandings of electric circuits are based on the interaction between their understandings of electric current and circuit connections; children may also construct a model of unequal electric current.

Shipstone's (1985) findings also indicate that children use different electric current models to explain different electric circuits; the nature of the electric circuit task influences children's thinking about electric circuits. This finding suggests that classroom assessments must contain multiple and diverse tasks and types of electric circuits in order to determine children's understandings. Along the same lines, Shepardson and Moje (1994) found that children's understandings of electric circuits are also based on their understandings of electric circuit connections and their procedural knowledge for connecting batteries, wires, and bulbs. This implies that children can successfully identify simple, series, and parallel circuits based upon the electrical connections. They may connect batteries, wires, and bulbs in complete circuits but without a scientific understanding of electric current. Assessment tasks that simply require students to draw electric circuits, predict if an electric circuit will work, or build an electric circuit may not accurately reflect children's understanding of electric current. Assessment task must also require students to explain, orally or in writing, why the electric circuit will or will not work.

Assessment for the Purpose of Diagnosing Children's Understandings

To diagnose children's ideas and engage them in the electric circuit unit, we asked children to predict and explain if the bulb(s) in the electric circuit diagrams would light or not. In addition, we revised the prediction task to include examples of series and parallel circuits, not included in the initial assessment task. This revision provided evidence of children's understandings about series and parallel circuits prior to instruction, and when administered at the end of the unit, provided evidence of change in children's understandings. We created a checklist based on the procedural and conceptual ideas necessary for understanding electric circuits (Table 4). The research literature guided the development of the checklist to ensure that we included the different conceptions children hold about electric circuits. The checklist served as the tool for assessing children's electric circuit predictions and explanations. An important aspect of this assessment was that children would also test their predictions in future activities. Thus, the activity was not only an assessment task, but also an instructional task to engage students in learning about electric circuits: a pedagogical-assessment activity. In this way, the assessment task was embedded in instruction.

Table 4. Checklist for Assessing Children's Electric Circuit
Predictions and Explanations

Checklist Category	Observation (Check appropriate rows for each category)
Prediction	
Accurate prediction	
Inaccurate prediction	
Procedural Understandings	
Accurate procedural understanding	
Inaccurate procedural understanding	
Understanding of Electric Current	
Explanation reflects one directional flow of electricity	
Explanation reflects equal clashing currents	
Explanation reflects unequal clashing currents	
Explanation reflects one directional flow of electricity with electricity consumed by the bulb	
Explanation reflects scientific perspective on electric current	

Assessments for the Purpose of Determining Children's Progress

To assess children's progress, Laura administered three assessments to correspond with the content being taught: simple circuits, series circuits, and parallel circuits. These assessments consisted of two tasks each: a draw-and-explain task completed by each child and a group practical task where each group would build and explain an electric circuit. Laura felt these assessments would provide a better picture of what her students knew about electric circuit connections and current at key points throughout the unit:

> I think it makes sense . . . it's a natural to test students after each type of circuit. I can see where this would provide more . . . give me a better idea about what they [students] have learned along the way. . . . I should have thought of this myself.

Several issues that we talked about before Laura decided on these assessments were time and management, assessment of individuals versus groups, and assessment of processes versus products. These issues illustrate the complexity of classroom assessment; that is, teacher assessment practice is not based solely on decisions about the type of assessment tasks to use or their knowledge about assessment. Laura had no difficulty with taking instructional time to implement the assessments, but she was concerned about the time it would require her to "grade" the additional science assessments in addition to the students' work in the other subjects:

> Well taking class time doesn't bother me . . . I must admit I am concerned about the extra time it's going to take to grade three more tests. . . you know there are other things to grade and do.

To address the time conflict, Laura decided that she would limit the draw-and-explain task to three electric circuit examples and one explanation that explained all three drawings. Although Laura wanted to have each student build a circuit, she was concerned about being able to observe each student and manage the class at the same time:

> It would be nice to have each child make their own circuit, but I just don't see how I can do it and observe them and still help others . . . make sure everyone is doing things. . . . I'll try a group observation, but that bothers me too . . .

Laura decided to use a group-based assessment task for building the electric circuits. At the same time, she had an internal conflict about giving everyone in the group the same grade:

> Giving a group grade is a problem for me and frankly for some parents. I am always concerned that some students don't do their fair share . . . they really haven't learned anything . . . their grade isn't fair.

The individual draw-and-explain task helped ease Laura's concerns about a group grade because it provided for some individual accountability. Laura also decided that she would observe each group building the circuit, although she was apprehensive about this. Part of her apprehension was based on managing the class and keeping the other groups on task while she observed one group's performance. By observing the group, she could assess the process of building a circuit as well as the product. She could also monitor each child's contribution to the task, easing her concerns about children not contributing.

To assist Laura in observing children's performance during the practical task, the observational checklist in Figure 1 was constructed. The checklist emphasizes children's procedural understanding, cooperation and contribution, and explanation or conceptual understanding. After much conversation about how to use the observational checklist, Laura decided on the following procedure: the group would receive one rating based on their performance in constructing the circuit; each child would receive the same rating; one child would be randomly selected to explain how the group connected the circuit, with each child receiving the same rating; one child would be selected to explain why the circuit worked with each child receiving the same rating; and each child would individually be rated on their cooperation and contribution. Each group member was encouraged to assist other group members in understanding how and why the circuit worked. In this way, the assessment task contained both individual accountability and positive interdependence. Each group was made aware of the performance categories on the observational checklist and how they would be rated so that all children knew what the expectations were prior to the assessment.

Laura decided to allow each group 10 minutes to complete the task, so she could observe each group during the class period. This aspect of Laura's electric circuit assessment plan was perhaps the most challenging because it required more classroom management and brought into opposition issues concerning individual versus group assessment, and process versus product assessment. These assessments, however, provided Laura with a better picture of student progress, as well as informed her about teaching. In her words, "I can honestly say I feel I know more what my students learned . . . although the group assessment still concerns me."

Circuit Type: _____

Group: _____   Date: _____

Performance Category	Students			
Group can connect battery, bulbs, and wires to construct circuit (simple, series, parallel)	S SD US	S SD US	S SD US	S SD US
Can explain how the group connected the battery, bulb, and wires to build the circuit.	S SD US	S SD US	S SD US	S SD US
Can explain why the circuit works.	S SD US	S SD US	S SD US	S SD US
Cooperates with others in completing the task.	S SD US	S SD US	S SD US	S SD US
Contributes ideas to the group for completing the task.	S SD US	S SD US	S SD US	S SD US
Score				

S = Satisfactory performance, 2 pts.
SD = Some difficulty, 1 pt.
US = Unsatisfactory performance, 0 pts.

Figure 1. Observational Checklist for Group Assessment

Assessment for the Purpose of Determining What Children Know and Can Do

The summative assessment for the electric circuit unit consisted of three tasks: a practical task, the electric circuit mystery box, a more traditional quiz, and the predict-and-explain task (Figure 2) administered at the start of the unit (to assess change). The predict-and-explain task required students to predict if the bulb(s) in the electric circuit diagrams would light or not and to explain their prediction. As a result of the development and use of the electric circuit mystery box task, Laura modified her initial instruction by discarding the final activity in which the children constructed an electric circuit blueprint. She felt the electric circuit mystery box assessment accomplished the same objective but in a more formal and structured way. Laura viewed the electric circuit mystery box task as both an assessment activity and an instructional activity. It also illustrates how she changed her instruction based on her assessment plan, an often-overlooked aspect in changing assessment practice.

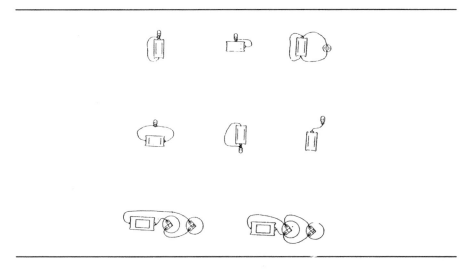

Figure 2. Laura's Predict-and-Explain Task

The electric circuit mystery box assessment task involved three phases. The first phase required each small group of students to plan (i.e., draw and explain) an electric circuit mystery box using a shoebox, wires, and four bulb holders and bulbs. The mystery box design was to include a simple, sires, and a parallel circuit. Before the students could proceed to the second phase, their plan had to be approved by Laura. The first phase was assessed based on the degree of assistance Laura provided each group and the accuracy of the group's explanations; therefore, there was an opportunity for the students to learn as they moved through the task. The second phase required each small group to build and test their electric circuit mystery box based on their plan. The third phase required the groups to exchange electric circuit mystery boxes and to identify and explain which bulb holders were wired in simple, sires, and parallel circuit. Although the children worked in groups, each child was individually responsible for completing the assessment, turning in a drawing that illustrated the three circuit types and supporting explanations. The third phase was perhaps the most difficult because of the number of possible combinations of circuit paths. The children were somewhat frustrated with this but generally worked well. The task was challenging but not overly difficult, so that children would not fail or lose interest. The level of challenge of a task depends on both the difficulty of the task itself and on students' perceptions of their ability to perform or complete the task (Natriello & Dornbusch, 1984).

---

1. Use the words in parentheses to describe and explain the electric circuit illustrated in the diagram. (Words to use: electric current, battery, circuit, conductor, wire, bulb)

2. The bulbs in the circuit do not light. Explain how you would change the wiring so that the bulbs will work.

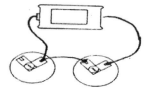

3. Timmy (a fourth grader) thinks the electric circuit illustrated in the drawing is a series circuit because bulb number three does not light. Mary thinks it is a parallel circuit because of the way the bulbs are wired; she thinks that bulb three is just "burnt out." Who do you think is right and why?

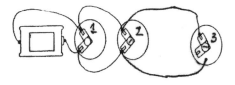

Figure 3. Laura's Revised Electric Circuit Quiz

Finally, Laura's quiz was changed to reflect three short-answer questions based on Bloom's (1956) taxonomy (Figure 3). This provided a variety of assessment techniques to determine students' science learning; all required students to use higher-order thinking (AAAS, 1998). The first question required children to use the language of science (e.g., "electric current," "battery," "circuit," "conductor," "wire," and "bulb") to describe and explain a simple electric circuit diagram. The second question required children to solve a problem they were to fix an electric circuit so that a bulb in the circuit would light. They then explained how they had solved the problem. The last question asked children to make a judgment and use evidence to support the judgement. Based on Bloom's taxonomy, the first question required students to apply their understandings of concepts and terms to an electric circuit diagram; the second question required students to analyze the electric circuit diagram to separate it into its essential elements or parts. The student had to recognize and discuss the relationship between the battery, bulb and bulb-holders, wires, and electric current in a series circuit. The third question reflected Bloom's evaluation category because the students had to use the evidence presented in the scenario as well as their knowledge about series and parallel circuits to evaluate Timmy and Mary's thinking. An important aspect of the second and third questions was the requirement to explain the answer.

EVALUATION AND GRADING

For Laura, evaluation was synonymous with grading and she encountered no conflicts in evaluating students (with the exception of her concern about group-based grades). Evaluation, however, is the process of drawing conclusions about student performance based on the assessment data collected, whereas grading is the assignment of a letter, numerical score, or percentage to the student or the student's work. In Laura's classroom each assessment task resulted in numerical scores that were then totaled to determine student grades. Each assessment task contributed proportionately to student grades. Although this approach is appropriate as a matter of teacher prerogative, it raises questions about the proportion or percentage that different assessment tasks should contribute to a student's grade. Most essentially: should the assessment tasks be weighted? If so how? Should assessment tasks be weighted by the type of student performance assessed? Should they be weighted by the type of assessment? A simple algorithm cannot answer this question; it should be answered based on what is valued in the science classroom. One way in which Laura could have weighted the electric circuit assessments is by assessment task, as follows:

Predict-and-explain (change from pre to post assessment)	15%
Draw-and-explain (simple, series, parallel circuits)	30%
Group-based circuit building (simple, series, parallel circuits)	15%
Practical task (electric circuit mystery box)	25%
Electric circuit quiz	15%

This weighting scheme places less emphasis on group-based assessments (a concern of Laura's) and greater emphasis on individual understanding (draw-and-explain). It also values change in children's understandings as much as their final quiz. The weighting scheme reflects the teacher's beliefs about the most import way of assessing student performance and values about science learning.

ASSESSMENT AS PROFILING

Although our conversations with Laura did not center on assessment as profiling, her assessment process for the electric circuit unit lends itself to thinking about this issue. Profiling is an approach that provides a system of assessment tasks and formats, these enable the teacher to construct a picture of student ability and progress based upon a variety of student traits or science domains. Profiling is not an assessment task, but a means of organizing and reporting student performance based on a variety of assessment tasks. The profile reports on similar student traits or science domains across the curriculum, illustrating a student's performance over time. It is a means to go beyond the traditional grade book, yet complementing the grade book. Assessment as profiling is in alignment with the National Science Education Standards Teaching Standard E (NRC, 1996) that calls for teachers to report student achievement in ways that go beyond grades. Based on work done in Britain (Balogh, 1982; ILEA, 1984), a good profile in science would:

- Record the assessment of the following domains of science: science process and inquiry skills, factual and conceptual understanding, problem-solving ability, cooperative learning ability and skills, and science attitude.
- Record the assessment of the following personal domains: motivation and commitment, responsibility, perseverance, and self-confidence.
- Report students' progress on the science and personal domains over time from the beginning to the end of the academic year. It is not necessary that every assessment task measure each domain, but that over time each domain be assessed several times and recorded on the profile.
- Be presented in a structured form and contain the same information for each student.
- Be made public to both students and parents.

Assessment as profiling requires us to think differently about what is assessed and about what we want students to know and be able to do in science. It requires us to think about which science domains are valued and how those domains may be assessed across the science curriculum. For example, the assessment profile depicted in Table 5 requires that instruction and assessment link to those domains, providing a picture of student performance within a domain across the science curriculum. For Laura, the assessment profile would require her to ask these kinds of questions:

- How do the electric circuit assessment tasks link to the domain categories of the assessment profile?
- How do students demonstrate conceptual understanding of electric circuits by using concepts to explain the results of an investigation?
- Do students plan and conduct an investigation of electric circuits?

Table 5. Example Assessment Profile

Student:

Science Performance Categories	Plant Unit	Weather Unit	Electricity Unit
Demonstrates conceptual understanding by using concepts to explain the results of investigations.			
Able to plan and conduct investigations and experiments.			
Cooperates with others in completing the task or assignment.			
Can make detailed and accurate drawings of phenomena.			
Able to obtain and use factual information from reference sources.			
Comments, Plant Unit:			
Comments, Weather Unit:			
Comments, Electricity Unit:			

Rating Scheme: S = satisfactorily, SA = satisfactorily with assistance, U = unsatisfactorily, NA = not assessed.

An assessment profile helps organize and focus the classroom assessment process across the curriculum. It helps to connect assessment tasks across the curriculum based on key domains of science learning instead of isolating individual and content dependent assessments.

Science assessment viewed as a kind of profiling would provide students, parents, and others with more detailed information about student performance and progress than do traditional grades or state-mandated tests. Such profiles would complement the traditional grade-reporting system by providing a more detailed basis for student grades. Still, the profiling process does require more teacher time and commitment, and it requires that an assessment plan articulate assessment and instruction across the curriculum. The intent is not to assess students in every science domain every time; this would be an overwhelming and unmanageable task. Instead, it is meant to sample student performance over time by establishing a sampling system (Lunetta, Hofstein & Giddings, 1981). A sampling system based on the sample profile presented in Table 5 might assess student performance on the use of concepts to explain the results of investigations in the weather and electricity units; it might assess student abilities to make detailed and accurate drawings of phenomena in the plant and electricity units, and it might consider student abilities to cooperate with others in completing the task in the plant and electricity units.

ASSESSMENT ISSUES AND CONCLUDING THOUGHTS

Laura's case brings to light various issues surrounding student opportunity to learn. The electric circuit mystery box task exemplifies an assessment task that provides students with the opportunity to learn. Laura assessed the task based on the students' plan and the degree of assistance needed to reach the goal of the task: to design a mystery box containing simple, series, and parallel circuits. Students were provided with an opportunity to learn and/or relearn about electric circuits as they progressed through and completed the task with appropriate teacher support and guidance. Students could not proceed to the next phase until they successfully completed the first phase. This view of the opportunity to learn through the completion of an assessment task is slightly different from that portrayed by the *National Science Education Standards* (NRC, 1996) which base the opportunity to learn on powerful indicators of science learning and on instructional experiences that provide students with opportunities to learn science.

Although the electric circuit assessment tasks described in this chapter and implemented in Laura's classroom reflected individual and group-based assessments that Laura scored, peer and self-assessments need to be considered and incorporated into classroom science assessment practice. This aligns the assessment process with instruction, as in this class, students are often required to collaborate on the completion of science activities; this should furnish opportunities to conduct peer assessments and self-assessments of collaborative and individual work. Student self-assessment engages students in reflecting upon their own performance to promote responsibility for learning and to build self-confidence. Peer assessments provide feedback about group work and the quality of academic work, and they establish a level of individual accountability. Peer and self-assessment may also inform pedagogy, providing teachers with insight into students' understandings and abilities. The *National Science Education Standards* (NRC, 1996) paint a picture of student self-assessment in which the teacher assists students in assessing and reflecting on their own science learning by formulating strategies and tasks that guide the students through peer and self-assessment.

In sum, through this collaborative dialogue I came to better understand Laura's classroom assessment practice, and Laura constructed a new understanding about assessment. This process was built on trust and respect, most importantly on my respect of Laura's decisions about her assessment practice and the reasons for her decisions. Although Laura was interested in learning more about assessment, it is unlikely that her electric circuit assessment tasks would have changed without collaborative dialogue and self-reflection on her current practice. It would be difficult for professional development programs to provide extensive one-to-one, instructor-teacher support as described in this chapter; however, schools can provide teachers with opportunities to learn about assessment by supporting teacher engagement in collaborative dialogue with colleagues about their classroom assessment practices. An important aspect of Laura's case is that she was interested in learning about new ways of assessing her students; she was open to ideas and ways of changing her practice, though initially she just wanted procedures to better assess her students' learning. The issue is how does one motivate and engage all teachers to change or improve their science assessment practices? Although assessment has received more attention in current reform documents, actual attention to assessment practice does not appear to be a priority at the district, school, or the classroom level.

It should also be noted that changing Laura's electric circuit assessments was very time intensive and evolved over several weeks. Although Laura has changed her assessments, she has not made wholesale changes in her classroom assessment practice: "I hate to admit it . . . I just haven't had-put in the time to think about assessment in other units." The extent of Laura's change was constrained by the time-space boundaries of our collaborative interactions and activity. This implies that, because of the time intensity of changing assessment practice, teachers are not likely to change their assessment practice beyond the time and support boundaries established by professional development programs. School and district level administrators who want teachers to change their classroom assessment practice must make classroom assessment a priority; they must commit to long-term professional development and provide ongoing support to teachers in both time and resources.

REFERENCES

American Association for the Advancement of Science (1998). *Blueprints for reform: Science, mathematics, and technology education*. New York: Oxford University Press.

Balogh, J. (1982). *Profile reports for school-leavers*. New York: Schools Council/Longman.

Bloom, B.S. (1956). *A taxonomy of educational objectives: Handbook I, the cognitive domain*. New York: Longman.

ILEA (1984). *Improving secondary schools: the report of the hargreaves committee*. London, England: ILEA.

Lunetta, V., Hofstein, A., & Giddings, G. (1981). Evaluating science laboratory skills. *The Science Teacher*, 48(1), 22-25.

National Research Council (1996). *National science education standards*. Washington, DC: National Academy Press.

Natriello, G. & Dornbusch, S.M. (1984). *Teacher evaluative standards and student effort*. New York: Longman.

Osborne, R. (1983). Modifying children's ideas about electric current. *Research in Science and Technological Education,* 1(1), 73-82.

Shepardson, D.P & Moje, E.B. (1994). The nature of fourth graders' understandings of electric circuits. *Science Education*, 78(5), 489-514.

Shipstone, D.M. (1985). A study of children's understanding of electricity in simple DC circuits. *European Journal of Science Education*, 6(2), 185-198.

Solomon, J. (1985). The pupil's view of electricity. *European Journal of Science Education*, 3(3), 281-294.

DANIEL P. SHEPARDSON AND EDITH S. GUMMER

6. A FRAMEWORK FOR THINKING ABOUT AND PLANNING CLASSROOM ASSESSMENTS IN SCIENCE

Which science domains should be assessed in order to determine what students have learned? What should be assessed within these domains? This chapter will focus on providing answers to these questions. The purpose is not to identify specific indicators or criteria for student performance because these would be specific to each individual assessment task; instead we describe the domains and performance categories that should be considered in developing a school district, school, or classroom-based assessment system. We present these assessment domains and performance categories to stimulate reflection about assessment practice and to provoke thought and dialogue about the development of assessment tasks at the classroom, school, and district level. Our assessment framework looks specifically at science content or discipline-based knowledge, science processes and inquiry skills, reasoning processes, and meta-cognitive processes. Although this chapter does not discuss science attitude and social/personal abilities, these too, should be considered in the classroom assessment system. These topics, especially science attitude, are described in other chapters in this book. After presenting our assessment framework, we will depict a professional development scenario that applies our assessment framework.

We first present an overview of changing assessment according to the *National Science Education Standards* (National Research Council [NRC], 1996). We then present two prominent science assessment frameworks: the National Assessment of Educational Progress (NAEP) and the Third International Mathematics and Science Study (TIMSS). Each includes domains and performance categories similar to those identified in our assessment framework. We acknowledge that some individuals may question the implementation of these frameworks; however, we believe they provide a perspective about assessment in science that should be reviewed and considered by anyone planning classroom-based assessments.

THE NATIONAL SCIENCE EDUCATION STANDARDS CHANGING EMPHASES OF ASSESSMENT

The *National Science Education Standards* (NRC, 1996, p. 100) presents a vision of assessment that reflects:

83

Daniel P. Shepardson (ed.), Assessment in Science, 83—97.
© 2001 *Kluwer Academic Publishers. Printed in the Netherlands.*

- Assessing what is more highly valued.
- Assessing rich, well-structured knowledge.
- Assessing scientific understanding and reasoning.
- Assessing to learn what students do understand.
- Assessing achievement and opportunity to learn.
- Students engaged in ongoing assessment of their work and that of others.
- Teachers involved in the development of external assessments.

What is valued in learning science is identified in the NRC (1996) Standards and is stated in Assessment Standard B: the ability to inquire; knowing and understanding scientific facts, concepts, principles, laws, and theories; the ability to reason scientifically; the ability to use science to make personal decisions and to take positions on societal issues; and the ability to communicate effectively about science.

The NRC states that, "Rather than checking whether students have memorized certain items of information, assessments need to probe for students' understandings, reasoning, and the utilization of knowledge" (NRC, 1996, p. 82). Further, the NRC calls for assessments that engage students in inquiry, that require students to use science knowledge to make personal decisions and to take and support positions on societal issues, and for students to communicate science to others. These changing emphases in assessment call for new frameworks to plan and implement classroom-based science assessments.

NATIONAL FRAMEWORKS FOR ASSESSING STUDENTS IN SCIENCE

National Assessment of Education Progress

Perhaps the most widely known science assessment framework in the United States is the National Assessment of Education Progress. A planning subcommittee representing the science education community was established to include science teachers, experts in science education, and scientists. The committee was charged with drafting a framework that would identify the most important outcomes of school science. This framework and the resulting assessment were to have the following five characteristics.

1. The framework should reflect the best thinking about the knowledge, skills, and competencies needed for a high degree of scientific understanding among all students in the United States. Accordingly, it should:

- Encompass knowledge and use of organized factual information, relationships among concepts, major ideas unifying the sciences, and thinking and laboratory skills.

- Be based on current understandings from research on teaching, learning, and student performance in science.

2. Both the framework and the NAEP science assessment should:

- Address the nature and practices of knowing in science, as different from other ways of knowing.
- Reflect the quantitative aspects of science as well as the concepts of the life, earth, and physical sciences.

3. Assessment formats should be consistent with the objectives being assessed. A variety of strategies for assessing student performance are advocated, including:

- Performance tasks that allow students to manipulate physical objects and draw scientific understandings from the materials before them.
- Open-ended items that provide insights into students' levels of understanding and ability to communicate in the sciences, as well as their ability to generate, rather than simply recognize information related to scientific concepts and their interconnections.
- Collections of student work over time (such as portfolios) that demonstrate what students can achieve outside the time constraints of a standardized assessment situation.
- Multiple-choice items that probe students' conceptual understanding and ability to connect ideas in a scientifically sound way.

4. The assessment should contain a broad enough range of items at different levels of proficiency for identifying three achievement levels for each grade.

5. Information on students' demographic and other background characteristics should be collected. Additional information should be collected from students, teachers, and administrators about instructional programs and delivery systems, so that their relationships with student achievement can be ascertained and used to inform program and policy decisions.

The committee endorsed curricula recommendations made by government agencies and professional societies (e.g., National Science Board, American Association for the Advancement of Science, National Science Teachers Association, and the National Research Council). These recommendations included:

- Reduce the traditional breadth of coverage of science concepts to allow for greater depth.
- Emphasize the development of thinking skills.
- A multi- or interdisciplinary approach to science teaching.

- Curricular and instructional approaches that encourage active student involvement.
- Increase participation of underrepresented populations in school science.

New to the NAEP assessment was the consideration of science themes that represent key organizing concepts cutting across the boundaries of science disciplines. Additionally, the assessment changed the category of "knowing science" to "conceptual understanding of science," substituted "scientific investigation" for "conducting inquiries," and changed "solving problems" to "practical reasoning." The performance tasks that probe students' abilities to use materials to make observations, perform investigations, evaluate experimental results, and apply problem-solving skills were also new to the assessment (O'Sullivan, Reese, & Mazzeo, 1997). These changes in the format of NAEP have not been without controversy.

The core of the NAEP framework is its two-dimensional organization. The first dimension addresses the earth, physical, and life sciences. The second dimension defines the characteristics of knowing and doing science: conceptual understanding, scientific investigation, and practical reasoning. Further, the framework includes the historical development of science and technology, the habits of mind, the methods of inquiry and problem solving, and the themes of science (e.g., systems, models, and patterns). The NAEP science assessment format includes multiple-choice, constructed-response, and hands-tasks (O'Sullivan, Reese, & Mazzeo, 1997, p.1):

- Multiple-choice questions that assess students' knowledge of important facts and concepts and that probe their analytical reasoning skills.
- Constructed-response questions that explore students' abilities to explain, integrate, apply, reason about, plan, design, evaluate, and communicate scientific information.
- Hands-on tasks that probe students' abilities to use materials to make observations, perform investigations, evaluate experimental results, and apply problem-solving skills.

Third International Mathematics and Science Study

The Third International Mathematics and Science Study (TIMSS) took a slightly different approach to determining the dimensions of the domains of science and mathematics analyzed by its achievement tests. Martin (1996) describes the overarching purpose of TIMSS as measuring student achievement or learning along with the context that supports that achievement. Before the TIMSS achievement tests were developed, an extensive analysis of the curricular documents and textbooks used in all of the participating countries was performed. This analysis resulted in a curriculum framework that provided the basis for the achievement tests.

The TIMSS curriculum framework evolved from a content by cognitive-behavior matrix that connected three dimensions (Martin, 1996): subject matter content, performance expectations, and perspective or context.   Subject matter content included the following dimensions:

- Earth sciences
- Life sciences
- Physical sciences
- Science, technology, mathematics
- History of science and technology
- Environmental issues
- Nature of science
- Science and other disciplines

Performance expectations described the performances that students are to demonstrate for a particular curricular unit.   For science these performance expectations included the following:

- Understanding
- Theorizing, analyzing, and solving problems
- Using tools, routine procedures, and science processes
- Investigating the natural world
- Communicating

Finally, perspectives or context described the curricular expression or dispositions particular to each science domain. These included attitudes, careers, participation, increasing interest, safety, and habits of mind.

TIMSS achievement tests measured both mathematics and science achievement in the same exam. Test format included both multiple-choice and free response items. Free response items included both short answer items that require students to generate a brief written response, and extended response items that require students to develop a more extensive written response that includes an explanation of their reasoning. Extended response items included performance assessment tasks that were designed to elicit indicators of students' responses to hands-on tasks organized into a circuit of stations.

A SCIENCE ASSESSMENT FRAMEWORK

The assessment framework we describe below both builds and expands on existing science assessment frameworks to provide a perspective on assessment in science.   The assessment framework also incorporates the national reform documents (e.g., *National Science Education Standards* and the *Benchmarks for Science Literacy*). Our framework is organized by assessment domains that are composed of performance categories with specific indicators or criteria linked to each performance category based on the assessment task and instructional context.

Because of the contextual nature of assessments we do not present indicators or criteria for the performance categories.

The Assessment Domains and Performance Categories

The assessment domains presented here are not necessarily exclusive to science or to science classrooms; they are based on the science education, assessment, and education literature. Although some may find our assessment domains to be somewhat limited, we believe they provide a starting point for teachers and administrators in the development of an assessment framework for their classroom, school, and district. We do not intend every classroom assessment task to address student performance in each domain; instead we propose that assessment tasks address multiple domains of students' science learning and performance capabilities throughout the academic year and across grade levels.

For each domain we share possible performance categories that may be considered in the development of assessment tasks. Again, we do not present specific indicators or criteria for the performance categories because these would need to be aligned with the particular assessment task and instructional context, and with the curricular goals for the classroom, school, or district. Our framework for thinking about assessment is presented in the form of a planning template shown in Table 1. From this template the actual assessment task and scoring system may be developed.

Table 1. Assessment Planning Template

Assessment Domain	Performance Category	Indicator or Criteria

Discipline-based Knowledge

Discipline-based knowledge is comprised of factual, conceptual, and procedural understandings. The standards for discipline-based knowledge have been well articulated in the *National Science Education Standards* (NRC, 1996) and the *Benchmarks for Science Literacy* (American Association for the Advancement of Science [AAAS], 1993) and we will not address them further. Teachers, however, need to be familiar with how this knowledge is articulated in these two standards documents. In addition, teachers need to determine how well the knowledge presented in textbooks and other curricular materials matches the standards. In other words, teachers need to be knowledgeable about how state content frameworks align with the national standards.

Science Processes and Inquiry Skills

The science processes that should be considered in assessing students' science performance include:

- Observing
- Classifying
- Measuring
- Inferring
- Predicting
- Explaining
- Experimenting
- Interpreting data
- Controlling variables

These process skills are derived from "Science A Process Approach" (AAAS, 1968), and have been well documented as essential to science and scientific inquiry, as well as reflect the dispositions to conduct scientific inquiry. The NRC (1996) Standards also stress that the process skills be developed in the context of inquiry and should be assessed in this context. The ability to inquire requires that students conceptualize, plan, and perform investigations (NRC, 1996). The *National Science Education Standards* (NRC, 1996) articulate the characteristics of inquiry at all grade levels to include:

> . . . asking questions, planning and conducting investigations, using appropriate tools and techniques to gather data, thinking critically and logically about relationships between evidence and explanations, constructing and analyzing alternative explanations and communicating scientific arguments (p. 105).

These characteristics go beyond individual science process skills to students' ability to conduct scientific inquiry. The NRC (1996) Standards also provide a more detailed description of the developmentally appropriate abilities for scientific inquiry for each grade level. These inquiry abilities can serve as performance categories for assessing students with specific indicators or criteria that reflect the context of the assessment task and the instruction. Assessments should also reflect the changing emphases to promote inquiry (NRC, 1996, p. 113), particularly:

- Extended investigations.
- Use of multiple process skills—manipulation, cognitive, procedural.
- Use of evidence and strategies for developing or revising an explanation.
- Application of results of experiments to scientific arguments and explanations.

The *Benchmarks for Science Literacy* (AAAS, 1993) may contribute to the development of assessment tasks for determining students' inquiry capabilities by providing performance indicators or criteria by grade level.   For example, a kindergarten through Grade 2 inquiry benchmark indicates that students should know that: "Tools such as thermometers, magnifiers, rulers, or balances often give more information about things than can be obtained by just observing things without their help" (p. 10). The Benchmarks (AAAS, 1993) also describe "Habits of Mind" that indicate benchmarks for computation and estimation, manipulation and observation, and communication. These habits of mind may serve as categories for assessing students' performances in science with the specific benchmarks as indicators of performance.   Again, the criteria for these habits of mind must be specific to the assessment task, set in the context of instruction.

Reasoning Processes

By developing assessment tasks and scoring systems that capture the reasoning processes, science teachers assess not only how students engage in science (science processes and inquiry skills) and how students understand science (content knowledge), but also how students reason scientifically. These tasks assist the teacher gather evidence of how students reasoning processes enable them to make sense of their experiences and the assessment task. Developing assessment tasks and scoring systems to capture students' reasoning processes reflects the changing emphases of assessment presented in the NRC (1996) Standards. Assessing students reasoning processes, however, is perhaps the most difficult dimension because, although the reasoning processes are embedded in many assessment tasks, they are often not directly assessed nor does the final product reveal the reasoning students used. Students' reasoning abilities may be inferred from the products they produce during the assessment task; however, the assessment product may not be a direct measure of students' reasoning abilities. Thus assessment tasks and scoring systems need to directly assess students' reasoning abilities.

Although a plethora of reasoning processes is identified in the educational literature, the essential reasoning processes include analysis, comparison, interpretation, and evaluation (Quellmalz and Hoskyn, 1997).   These reflect cognitive strategies for processing and using data and information: *analysis* involves identifying the elements or parts of the whole and understanding the relationship between the parts and the whole. *Comparison* involves the identification of similarities and differences between phenomena and processes. *Interpretation* is the process of making sense of situations and results using existing understandings and knowledge. *Evaluation* is the process of making judgment about a situation or event based on evidence, understanding, and most importantly criteria. These essential reasoning processes may serve as performance categories within the reasoning domain.

The *Benchmarks for Science Literacy* (AAAS, 1993), "Habits of Mind," presents critical response skills that involve reasoning processes. "These critical-response skills can be learned and with practice can become a lifelong habit of mind" (AAAS, 1993, p. 298). Critical response skills include questioning claims, making judgements, supporting positions with evidence, and being skeptical of arguments. These skills may serve as performance categories for assessing students in science. The NRC (1996) Assessment Standards also indicate that scientific understanding includes the ability to reason with science knowledge and information, to make and justify predictions and to develop explanations.

Meta-cognitive Processes

Meta-cognitive processes are the self-conscious and deliberate ways individuals monitor their reasoning processes and activity as they solve problems, make decision, or complete tasks. Key meta-cognitive processes involve planning, monitoring and revising, and evaluating and reflecting (Quellmalz and Hoskyn, 1997). In science, meta-cognitive processes are used in planning and conducting investigations or experiments. Requiring students to monitor, evaluate, and revise their science investigations or experiments also reflects the use of meta-cognitive processes. Therefore, assessment tasks and scoring systems could capture students' abilities to use these meta-cognitive processes in science.

The *National Science Education Standards* (NRC, 1996) for inquiry also articulates a set of meta-cognitive processes: planning investigations and thinking critically and logically about relationships between evidence and explanations. These meta-cognitive processes may serve as performance categories in assessment tasks; however, like the reasoning processes, they are difficult to assess without deliberate planning.

ASSESSMENT PLANNING SCENARIO

The assessment planning scenario that follows illustrates one way that professional development could be delivered in a school or school district. The scenario describes how a science supervisor might use the assessment framework and the assessment planning template presented in this chapter, to involve teachers in thinking about and planning classroom-based assessments. The scenario also describes how teachers could use national and state standards documents as tools for thinking about and developing both instruction and assessment, demonstrating the inter-relatedness among curriculum, instruction, and assessment.

Dr. Jane Blom, the K-12 science supervisor for the Winndott School District in West Central Indiana, was preparing the third in a series of professional development activities for her 16 teachers.   She was planning a meeting for the three K-5 science specialists who were assigned to the three elementary schools in the district, six middle school science teachers and seven high school science teachers.   This year the teachers requested a focus on assessment for their five professional development meetings.   Previously they had continued an ongoing discussion of the NRC *Standards* and the AAAS *Benchmarks* that they had used to examine and modify their curriculum in keeping with the state level science standards and high school science competencies.   They had examined the presentation of science inquiry and the process of assessment in these documents.   Dr. Blom determined that the teachers needed a concrete example of an assessment task, one to which they could all relate and one that would examine the issues important to integrating assessment with instructional practices, curriculum development, and student learning over time.

The teachers shared the assumptions about important themes from the AAAS Benchmarks and the state's science standards. Dr. Blom knew that the assessment domains had to link with those themes. She looked again at the NRC Standards and decided to use the 'plant in the jar' task (NRC, 1996, p. 92) as an example of an assessment task that might parallel instruction. She presented the task to the group as follows:

> Some moist soil is placed inside a clear glass jar. A healthy green plant is planted in the soil. The cover is screwed on tightly. The jar is located in a window where it receives sunlight. Its temperature is maintained between 60° and 80°F. How long do you predict the plant will live? Write a justification supporting your prediction. Use relevant ideas from the life, physical, and earth science to make a prediction and justification. If you are unsure of a prediction, your justification should state that, and tell what information you would need to make a better prediction. You should know that there is not a single correct prediction (NRC, 1996, p. 92).

Dr. Blom then split the teachers into two groups and asked them to determine the assessment domains that the task fit (see the first column of the assessment planning template).   She then asked the groups to determine how they might use the task to assess students in both the second grade and the high school biology level. She asked for a description of the instructional activities and the performance categories and indicators or criteria the teachers would develop to assess students. (These would be displayed in the second and third columns of the template.)

Both groups looked in the AAAS Benchmarks and determined that the task addressed the living environment (science content) and the reasoning domains. The teachers then identified the performance categories related to the science concepts embedded in the task and the reasoning the task required. Group members rarely disagreed about either assessment domains or the general performance categories. The determination of the indicators or criteria, however, was much more difficult; multiple discussions were needed to examine the instructional activities and the science education standards documents.

The group that was examining the task for use in the second grade decided to use the prompt as a summative assessment after the students had completed a guided inquiry unit about terrariums. The students would have set up terrariums and made observations for six weeks as they examined the requirements for plants to grow and survive. Throughout the project, classroom teacher would stress the need for continual observation of the plants and the recording of those observations in students' journals. Students would also explore possible modifications of the terrariums in order to better determine what factors influence plant growth and survival; for example, some groups of students might withhold water or sunlight from the plants. The teacher could read a story that described how plants use the air to make food. The teacher would also elicit a brief writing description from the students to ascertain what they understood about how plants make their food. This would include a drawing that could reveal potential misconceptions for later classroom discussion.

After this instructional period, the teacher would demonstrate the construction of a small plant in a large jar with some soil and water. The teacher would seal the jar and place it near a window for the class to continue to observe. She would then present the 'plant in the jar' task to the students on a response sheet and ask them to work in small groups to verbally discuss a response to the task. She would circle among the groups listening to their responses and ask each group to come to a consensus that they would present in written form. The teacher would make observational notes about participation in the groups and would then evaluate the final responses that the students wrote. An exemplary response would address the indicators shown in Table 2 below.

In contrast, the high school biology group decided to use the 'plant in the jar' task in two ways; the first would involve a summative assessment after the students had explored the reciprocal processes of photosynthesis and respiration. The second use was as a formative assessment/instructional device to introduce the ecology unit into which the class was moving. In this case, the teacher would have needed to introduce the concept of systems early in the year so that students would be accustom to looking at biological systems in terms of inputs, outputs, and processes that occur within systems. One of the middle school life science teachers and the two biology teachers described various instructional activities that introduced students to systems.

Table 2. Assessment Planning Template Example for Second Grade

Assessment Domain	Performance Category	Indicator or Criteria
Science content	Living Environment: Interdependence of Living Things Flow of Matter and Energy	Explanation should relate to instructional activities that describe interdependence. Student responses should address the plant's need for water, sunlight, air, and soil.
Reasoning processes	Makes Prediction	Student should make a prediction that relates directly to the growth/survival of plant.
	Justifies Prediction	Justification should relate prediction to evidence from instructional experiences students had with terrariums and the science content.

Table 3. Assessment Planning Template Example for High School Biology

Assessment Domain	Performance Category	Indicator or Criteria
Science content	*Living Environment:* *Interdependence of Living Things* *Flow of Matter and Energy*	*Explanation should relate to instructional activities describing the reciprocal processes of photosynthesis and respiration. Student responses should describe how the plant's requirements for light as the energy source, H_2O, CO_2, and O_2 are being met. In addition, explanations should reflect instructional discussions about systems from earlier instructional activities.*
Reasoning processes	*Makes Prediction* *Justifies Prediction*	*Students should make a prediction that relates directly to the survival of the plant.* *Justification should identify assumptions made and science content, and should include information students identified as missing from the prompt.*

Before using the 'plant in the jar' task, the biology teacher would have to link her traditional laboratory experiences in photosynthesis and respiration to the notion of systems without explicitly describing the task. The teacher would then model the set-up and give the students a response sheet. The teacher decided to examine the students' abilities to make the connections individually, so no group work was included. The teacher would assess the individual responses and then return them to the students. An exemplary response would address the indicators shown in Table 3. The students would discuss their responses in small groups and explore in more detail the assumptions they had made while constructing these responses. A whole class discussion would identify these assumptions, highlighting their assumptions about aspects of the system besides just photosynthesis and respiration. The teacher would lead the students to discuss the 'plant in the jar' task in terms of an ecological system in order to identify other inputs, outputs, and processes that might be occurring. The students would then study other ecological systems through field trips and laboratory experiences.

In summary, this scenario described one way that a group of teachers determined what they needed to consider when planning an assessment system. The 'plant in the jar' task was open-ended and could be examined by students at different developmental levels. The responses given by the students would vary, but the comparison of students at the second grade level with high school students shows how central themes such as interdependence and the flow of matter and energy in the living environment can be articulated throughout K-12 school science curricula. These uses of the 'plant in the jar' task, both formative and summative assessment, demonstrate that the teachers were using the assessment data for more than grading purposes.   This scenario demonstrated the inter-relatedness of curriculum, instruction, and assessment, as well as the use of the national science education standards documents to plan assessment tasks.

CONCLUDING THOUGHTS

The intent of this chapter was to stimulate thinking about assessment domains and performance categories that might be used to assess students in science. The framework presented here is not exhaustive, but it does provide a basis for initiating dialogue and action toward developing a more comprehensive assessment plan for classrooms, schools, and districts. The framework is based on our collaboration with teachers to develop classroom assessments. We have learned that teachers have difficulty in developing assessment tasks that explicitly assess students' reasoning and meta-cognitive processes.

Both the developmental level and the collective instructional experience of students must be considered in the development of assessment tasks. These determine what constitutes appropriate assessment; for example, a third grade student's ability to analyze science data and conduct science investigations differ from that of a tenth grade student. For this reason, the specific indicators or criteria for each performance category must reflect the developmental level and instructional experiences of students.

REFERENCES

American Association for the Advancement of Science. (1968). *Science: a process approach.* Lexington, MA: Ginn.

American Association for the Advancement of Science. (1993). Benchmarks for Science Learning. New York: Oxford University Press.

Martin, M. O. (1996). *Third international mathematics and science study technical report volume 1.* Chestnut Hill, MA: TIMSS International Study Center, Center for the Study of Testing, Evaluation, and Educational Policy, School of Education, Boston College.

National Research Council. (1996). *National science education standards.* Washington, DC: National Academy Press.

O'Sullivan, C.Y., Reese, C.M., & Mazzeo, J. (1997). *NAEP 1996 science report card for the nation and the states: Findings from the national assessment of educational progress.* Washington, DC: Office of Educational Research and Improvement, U.S. Department of Education.

Quellmalz, E. & Hoskyn, J. (1997). Classroom assessment of reasoning strategies (pp. 103-130). In G.D. Phye (Ed.), *Handbook of classroom assessment: Learning, adjustment, and achievement* (pp. 103-130). San Diego, CA: Academic Press, Inc.

DANIEL P. SHEPARDSON

INTRODUCTION TO SECTIOIN II:
CLASSROOM-BASED ASSESSMENT TECHNIQUES AND
TEACHER CASE STUDIES OF ASSESSMENT PRACTICE

This section of the book (Chapters 7-15) presents practical, classroom-based assessment strategies and tasks that may be utilized by teachers in their classroom. Chapters 9 through 14 are case studies of teacher assessment practice written by teachers. In these teacher-chapters, the teachers tell their story about their assessment practice and how their assessment practice has evolved over time. The purpose of this section is to provide staff developers, administrators, and teachers with examples of practical assessment tasks and approaches that may be adapted for use at the district, school, or classroom level. Through the teacher stories, the section provides insight into the possibilities and limitations of classroom assessment in science.

The first chapter in this section (Chapter 7) articulates alternative approaches and techniques to assessing young children's science learning, integrating the teaching and assessing of children's science and literacy development. The chapter talks about the role of the teacher in assessing young children and cautions against restrictive teacher-child interactions that limit the potential of young children and bias our judgements of young children's performance. Chapter 8 describes specific techniques for assessing elementary and middle school students' science learning and abilities, including techniques for assessing students' attitudes, utilizing peer and self-assessments, and assessing students' self-produced journals. In Chapter 9, an elementary teacher tells her story about the assessment process she uses in her classroom. The teacher outlines how she develops and refines her assessment tasks, how she uses observation and interview techniques to assess students, and how she incorporates practical assessment tasks in a station format. The chapter also contrasts formal versus informal assessment.

Assessment tasks utilized in a middle school life science classroom are shared in Chapter 10. In this chapter the teacher tells her story about developing assessment tasks that serve as teaching-learning tasks, presenting and discussing samples of student work. The teacher describes how she used the *Benchmarks for Science Literacy* (AAAS, 1993) as a tool for informing her about children's ideas and performance capabilities. Chapter 11 continues the middle school focus, by discussing how one teacher used the *Benchmarks for Science Literacy* (AAAS, 1993) as a tool for developing standards-based assessment tasks and scoring guides for an end-of-year assessment instrument. The chapter includes example assessment

Daniel P. Shepardson (ed.), Assessment in Science, 99—100.
© *2001 Kluwer Academic Publishers. Printed in the Netherlands.*

tasks from the instrument and discusses student work samples. In Chapter 12, a Junior High School teacher tells his story about developing assessment tasks for the purpose of diagnosing students' misconceptions in an earth science unit. Example assessment tasks and student work samples are presented and discussed in light of the scoring guide and student conceptions.

A high school chemistry teacher, in Chapter 13, describes how she uses alternative teaching and assessing methods: computer animations and other forms of visualizations. The chapter illustrates how closely entwined teaching and assessing are in science classrooms. The teacher tells her story about how she developed, implemented, and revised her classroom instruction and assessment practice based on the assessment data she collected. The last teacher-chapter (Chapter 14) also tells a story about how a high school teacher's assessment practice evolved over time. The teacher articulates his perspective about authentic assessment and how he uses student peer and self-assessments. Examples of authentic assessment tasks are presented and the teacher's reflection and evaluation of those tasks.

SUSAN J. BRITSCH

7. ASSESSMENT FOR EMERGENT SCIENCE LITERACY IN CLASSROOMS FOR YOUNG CHILDREN

In early childhood classrooms from preschool through third grade, an emergent literacy approach defines literacy as activity, talk, listening, drawing, and writing (Shickendanz, 1986; Dyson, 1989; Morrow, 1997). All of these develop in situational contexts as young children interact with others for purposes that are both new and familiar. This means that emergent literacy does not view the development of oral and written language as imitation, but as an interaction in which "...the others are not simply providing a model—they are also actively engaged in the construction process" with the child (Halliday, 1980, p. 8). This means that in classrooms for young children, not only the children's language but also the teacher's language should be assessed for its continued development over time. Child language functions and teacher language functions differ in the context of science experiences as the teacher guides, questions, highlights contradictions, and helps the children to formulate their questions and problems so that *they* can answer and solve these.

Drawing from our observations of young children's science activity and emergent science literacy, we can develop assessment profiles to record children's science ideas along with their emergent literacy behaviors. Such profiles give attention to both science content and children's uses of emergent literacy for specific purposes. In this way, we can observe the specific forms that children use to conventionally and inventively record, interpret, and communicate science content. Science content assessment instruments add documentation of each child's prior and resultant knowledge about the phenomenon in question.

Different genres serving different functions (e.g., stories, lists, letters, science journals) may produce different oral and written language outcomes for different children. Therefore, literacy assessment criteria can be genre specific and tailored to well outlined aims for each experience. If the children are to engage in an experience that lends itself to the construction of a data table, for example, do the assessment criteria account for the literacy abilities necessary for this form? How will assessment value a story that the child draws or writes to describe science phenomena? This chapter will suggest some alternatives for the assessment of emergent science literacy, focusing on both the science ideas and the graphic (i.e., writing and drawing) forms in child-created science literacy products. These assessment tools chronicle and interpret children's science activity. To illustrate the

Daniel P. Shepardson (ed.), Assessment in Science, 101—117.
© 2001 *Kluwer Academic Publishers. Printed in the Netherlands.*

assessment of both the oral and graphic channels of child activity in science experiences, the next section presents a view of a kindergarten science experience with dissolving substances.

TALKING, DRAWING, WRITING: AN EXPERIENCE WITH
DISSOLVING SUBSTANCES

In one kindergarten, small groups of children investigated the topic of "dissolving substances" over the course of three days, for 30-45 minutes each day. The investigation began with a child-dictated list of definitions for the word "dissolve." The fact that things "dissolve in water" was most salient for many children; for others, dissolving was synonymous with "disappearing" or with "melting." One child, Nicholas, did not differentiate dissolving from dispersion, remembering a previous classroom activity in which the teacher had dripped eyedroppers of food coloring into a petri dish of water. Another child, Deanna, equated dissolving with "something white," recalling an earlier investigation of differences in the properties of salt, sugar, baking soda, and flour. Jeffrey said, "Water gets dissolved," when asked for his definition. He had reversed the process, which is consistent with the preoperational child's difficulty in thinking about the relationship of one event to another in a transformation (Brewer, 1995). These orally expressed ideas can be recorded on an assessment chart such as the one shown in Table 1.

In small groups, the children next tested the dissolving capacities of sugar, salt, soil and sand using chemplates and eyedroppers of water. The children summarized their results on a teacher-prepared data table that was stapled into each child's science journal. They then constructed more open-ended responses on a blank page where they could write or draw about what they had learned. Deanna's drawing, shown in Figure 1, depicted a duck walking across a beach. She explained:

> She doesn't like that dirt mixing up with the sand between her toes because it's too messy and it's dissolving.

Although Deanna's definition of dissolving had previously been linked only with white substances, soil and sand provided a new set of variables (i.e., color and texture) that became more salient for her through the small group task. She conveyed this on her journal page through her make-believe duck and her imaginative oral narrative.

Even after the dissolving activity Nicholas continued to equate dissolving with dispersion, drawing a petri dish and eyedropper labeled "i" (shown in Figure 2). He explained:

> Here's a bowl? And the food coloring from an eyedropper is dissolving and that's why I put a 'd' here.

Jeffrey created his own version of the teacher-provided data table that he had already completed as shown in Figure 3. Although he found the results for sugar, salt, and sand uninteresting, he was intrigued by the reaction of his soil sample, which contained some small pebbles. He recorded soil as a non-dissolving substance. Because the pebbles in his soil sample didn't dissolve, he invented his own term, "fracturing," to reflect a middle ground between dissolving and not dissolving:

> The dirt is *fracturing*. Some parts of it are on top and some are on the bottom. It's different from dissolving.

In summary, while Jeffrey used his journal to simply record results, he elaborated and interpreted them orally. Nicholas's entry still inaccurately analogized dissolving with dispersion. Deanna, however, invented an imaginary setting to contextualize her understanding of dissolving. Table 1 records both prior and current child understandings as part of the science content assessment for this experience. By allowing the teacher to review conceptual development across children, this sort of profile also points up issues in program continuity (such as the food coloring demonstration) that may impinge on child understanding.

Figure 1. Deanna's Duck   Figure 2. Nicholas's Petri Dish with
                                          Food Coloring

A similar profile, also illustrated in Table 1, accompanies the science content assessment to document the children's science journal use by describing the literacy forms used (e.g., drawing, copying, patterns of invented spelling). Jeffrey's journal entry and talk indicate that he grasped the phenomenon. Jeffrey recorded his results by copying written language and interpreting these results through oral language. Like Deanna, his literacy experiences might now include dictation to record his own highly accurate technical terminology on the page. This will begin to pull him away from copying to assure accuracy and toward symbolizing his own ideas in print.

Deanna developed a fictional narrative through oral language and drawing. She might now benefit from dictation experiences to record her imaginative narratives in print. In terms of science concepts, she will need further explorations to clarify whether she views soil and sand, or both, as dissolving substances—in other words, whether she understands the concept.

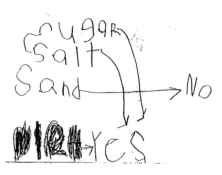

WHAT HAVE I LEARNED
ABOUT
Dissolve

Figure 3. Jeffrey's Data Table

Nicholas expressed the analogy he perceived by using oral language and by depicting a procedure through drawing, labeled with an attempt at representing a word-initial sound with a letter. He might now begin to expand his spelling attempts to focus on other letter-sound combinations beyond the word-initial ones. His conceptual equation of dissolving with dispersion was perceptually based and may be difficult to dislodge without greater understanding of the reasons that lie behind the observed phenomena. In the case of the dissolving experiment, a more difficult contradiction was posed for children whose prior understanding led them to explain that "dissolve" meant "disappear." When they dissolved sugar in water and then tasted the sugar water, the substance became invisible and seemed to disappear but could still be tasted. Such a contradiction was difficult for many young children to resolve conceptually. Because young children's learning is highly perceptual, curricular assessment must consider whether and how a set of experiences builds conceptual understanding or whether it simply results in a manipulative task to be completed.

Table 1. Assessment Profiles: Science Journal Use and Ideas about Dissolving

Child's Name	Drawing	Uses Invented Spelling	Copies Letters/Word
Jeffrey			X
Deanna	X		
Nicholas	X	*"i" = eyedropper*	X

Child's Name	Child's Prior Knowledge	Child's Current Knowledge
Jeffrey	*"Water gets dissolved"*	*Dirt is "fracturing"*
Deanna	*"Something white"*	*Teacher question: do both dirt and sand dissolve?*
Nicholas	*Food coloring dissolves in water*	*"Food coloring from an eyedropper is dissolving."*

Profiles such as those shown in Table 1 reveal patterns that assist teachers in designing small-group instruction that is tailored to the literacy and conceptual needs of the children. Table 2 records in a more detailed way the graphic forms used by a small group of kindergartners in their science journals. These criteria consider Clay's (1975) principles of beginning writing behavior as well as patterns in children's uses of invented spelling (e.g., Graves, 1983; Bear, Invernizzi, Templeton & Johnston, 1996). While the forms children use to produce written language are often emphasized in school assessments, emergent literacy views as equally important the oral and graphic processes children use to develop and convey understandings. To reflect this, Table 2 also includes the results of teacher-compiled observations characterizing the different processes children used to combine pictures and language use as they constructed science journal entries (Dyson, 1989).

As shown in Table 2, children like Ellen and Jacob convey information primarily through pictures that they complete before they talk about their understandings. For Jarrod, the meaning lay in the talk and he later expressed his understandings in writing and drawing. Children like Mollie wrote silently and then read back what has been encoded in print. Janet talked about her picture as she produced it, and this orally elaborated the details in the drawing. Such an assessment profile vividly illustrates the different processes used by the children in a group: two children first made their science ideas audible through talk; two others needed to begin by making their ideas visual through pictures; another began by expressing her ideas through print. Attempts to change the order of the child's process often result not only in a less elaborated product but also in less clearly expressed ideas (Dyson, 1989).

Table 2. Emergent Literacy Principles and Processes

Child Name	Invented Spelling	Flexibility Principle	Space Concept	Use of Drawing/ Written Language	Oral Language Use
Ellen	*Includes vowels*	*"p"/"b" confused*	*Accurate*	*Draws and writes*	*Pictures were drawn first. She said she had the words in her head. She wrote when she had finished drawing*
Jarrod	*Includes vowels*		*Accurate*	*Writes and draws*	*The journal entry was a dialogue. His writing provided an explanation.*
Jacob	*Word-initial and word-final consonants*	*"d" / "b" confused*	*Accurate*	*Writes and draws. Talks later.*	*The content was mostly in the pictures. He would not talk about his entry while drawing.*
Mollie	*Includes vowels.*		*Accurate*	*Writes and can reread.*	*She wrote the entry and then read it back. She did not talk while writing.*
Janet	*Not observed.*		*Not observed.*	*Draws with talk*	*She talked about details in the picture as she drew.*

Through their literacy capabilities, young children can contextualize the science experience on the page, bringing the most salient elements of that experience into this highly individualized frame. What children represent are their current explanations for phenomena that they contextualize using their own thinking, experiences, and imaginations. This does not always show the teacher the standard explanation for a phenomenon, but instead provides an individual point of view as the child connects the external, or visible, phenomenon with an internal, or invisible, world. To use science journals productively children must be able to pull the visible elements of the science experience into an internal world that is familiar to them. It is through the visible reflection of understanding on the page that teachers can assess each child's science ideas and science literacy capabilities.

TEACHER INTERACTION

As teachers, we need to monitor our own talk as well as the children's to better carry out teacher-small group interactions. As a result of such self-assessment, we may develop different kinds of questions, suggestions, and comments to involve and engage all students. Teachers may record effective and ineffective ways to elicit and build on children's ideas and understandings and promote the children's own questioning of ideas. We may also question how we can better chronicle the history of successive solution attempts, help children transfer this record to their journals. Is much of the children's science time spent simply asking for directions to carry out the procedure for the investigation? If so, how can teacher questioning be refocused to encourage the children's independent inductive reasoning about what to do? By monitoring our own ways of using language with children, we can more responsively design instruction, mediating children's interaction with others to encourage participation and collaboration. Through appropriate self-assessment, teachers can decenter their approaches to interaction, revising ways to alternately direct and be directed by the children's language in an appropriately designed scientific situation.

Table 3 offers some language strategies (Tough, 1979)--possible ways of using language--that teachers can note when reviewing their own talk. Other language strategies may also be valued by the teacher or weighted as more or less central to the effectiveness of ongoing classroom interactions. Table 3 also illustrates some features to note when reviewing the children's talk in the context of a daily science experience. A number of these teacher and child language strategies are the same, and reflect a sociocentric view of development in which "...the child's construction of reality...takes place through interaction with others" (Halliday, 1980, p. 16). In language learning at home as well as at school, these others also use language in particular ways for particular purposes in particular situations. Children acquire language by observing and participating in such situations; thus, the language used by significant others becomes language that children themselves use for their own purposes.

These kinds of language strategies can be described more individually and illustrated by examples on checklists, such as those shown in Tables 4 and 5, to record both milestones in learning and problems that require additional attention. Language strategies help us to follow and adapt the development of talk in the classroom, and not simply to evaluate "correct" versus "incorrect" responses. By using such documentation, teachers can ask themselves more useful questions: if children respond primarily with yes/no answers, is the teacher's talk offering them opportunities to justify their claims? Does the teacher's talk provide a cognitive model, leading children toward logical thinking and its articulation in talk? Through such assessment teaching aims, methods and outcomes can better coincide. Each teacher's checklist will contain his or her own criteria, reflecting that teacher's definition of productive interaction and thus the most salient values for assessment in that classroom.

Table 3. Some Classroom Language Strategies

Some Teacher Language Strategies	Some Child Language Strategies
Repeat child's statement	Give an example
Give instructions	Give a yes/no answer
Give factual information	No response
Direct child's action	Agree with teacher
Clarify child's response	Disagree with teacher
Ask yes/no question	Label an object, person, action
Ask labeling question	Describe an object, person, action
Ask prediction question	Describe an experience
As logical question	Describe a procedure for doing
Inductive	something Describe own actions
Deductive	Give factual information
Ask experiential question	Express use of inductive logic
Ask imaginative question	Express use of deductive logic
Propose a solution	Responds by using comparison
Collaborate with child to construct a	Responds by giving detail
solution	Anticipate a possibility for future
Use comparison	action
Describe details	Justify own claims
Anticipate a possibility for future	Criticize others
action	Propose an action
Justify own claims	Use language to guide own actions
Propose an action	
Use language to guide own actions	

CURRICULAR ASSESSMENT: EXPERIENCES WITH MAGNETS

Along with assessment of moment-to-moment interactions and daily literacy products, teachers can also focus their assessment on the appropriateness of the science phenomena to be explored by young children. We may clarify our aims in teaching and ask ourselves how much children at each developmental level can directly manipulate, discover, change, and understand. To illustrate, this section will examine several experiences with magnetism that are often provided for young children.

Table 4. Sample Checklist for Assessing Teacher Language Strategies

Child's Name	Leads to constructive Response	Leads to accurate factual information but a non-constructive response	Leads to imitative response or no response	Not Observed
Sam	*(date)* *Ask prediction question requiring inductive logic:* *"What if we take the leaf off the tree and we bring it in? Would it still work?"*			

Table 5. Sample Checklist for Assessing Child Language Strategies

Child's Name	Factually accurate information and a constructive response	Factually inaccurate information but a constructive response	Factually inaccurate information without revision	Not observed
Sam	*(date)* *Uses inductive logic to express cause-effect in tree growth.*	*(date)* *Proposes an action: "We could bring [the leaf] to school and show the class how it's turning green"*		
Nicholas	*(date)* *Anticipates possibility for future action: chiseling diamond and crystal*		*(date)* *Uses comparison: equates dissolving with dispersion.*	

A typical list of activities for a magnet unit might include the children's use of magnets to explore a provided set of objects (e.g., a penny, a ball of aluminium foil, a paper clip, iron filings) to test for attraction versus non-attraction. In his kindergarten, Nicholas did just that and recorded his findings in his science journal. His magnet page is shown in Figure 4.

Nicholas used the word-initial consonant, "M" to label the two magnets he drew. He surrounded each magnet with wispy lines that he labeled with "P" to signify, as he explained, "air pressure." He explained that air pressure made the magnets attract. When his teacher placed a magnet on the underside of the blackboard's chalk tray (non-attracting) and the magnet fell off, she asked him why the magnet didn't stick. He replied, "Because the wind isn't blowing that way." A number of the other children accepted and repeated the "air pressure" explanation, adopting this scientific-sounding language. As a next step, Nicholas might watch a fan pushing air or "wind" against the magnet on the underside of the chalk tray and see the magnet fall again. This would provide a cognitive and perceptual contradiction to his previous understanding, but this may not help him to comprehend the reasons behind the phenomenon he saw. Magnetic attraction requires the presence of iron in the substance. Kindergartners may be taught to repeat the explanation that "there is iron inside" but they must take this explanation on faith without the ability to see and touch that substance. In this case, Nicholas's teacher did not bring the fan; her aim was to provide the children with opportunities to observe *that* some objects attract a magnet and not *why* they do so.

Figure 4. Nicholas's "Air Pressure" Journal Entry

The same issue arose when kindergartners in another class applied magnets to different objects around the room. The children were to use their science journals to record which objects attracted and which did not attract the magnet. After the children observed the phenomena of attraction and non-attraction, they separated the items into these two categories on their journal pages. Several of the children recorded their classifications by simply indicating how many attracting and non-attracting objects they found, as in Figure 5, or by attempting to draw those objects, as in Figure 6. Because many of these kindergartners did not yet use detailed representational drawing, their journal entries were difficult to interpret without constant teacher interaction to clarify the children's ideas through talk. After the fact, the children were not able to recall what they had drawn or the objects that did attract. In their science journals, the children did not represent any explanation or conclusions for the observed phenomena; their understanding essentially stopped with the classification of found objects into two columns.

For very young children, magnetism is an invisible force whose physics can be observed but not conceptually understood. What, then, is the teacher's aim for magnet experiences? It may, in fact, be to provide children with opportunities to observe magnets with various objects accompanied by simple classification. In the above case, however, the children's literacy products were limited to recording the results of a procedure in which the children engaged without later opportunities to explain or interpret their results.

When asked to draw or write what they knew about magnets, most of the children in this kindergarten drew refrigerators on which magnets held up rectangles and squares (drawings and other pieces of paper), as in their own kitchens. These kinds of literacy examples on the page can lead teachers to ask how a unit about magnets can help young children solve problems in their daily lives (Gatzke, 1991). If the children's daily interaction with magnets is very limited, the utility of such science experiences can be questioned. This view of curricular assessment confronts the teacher with the necessity to distinguish between teaching that is merely activity-based versus teaching that is conceptually-based in educationally and developmentally appropriate ways. Gatzke councils that "developmental appropriateness" refers only to the match between ability level and teaching method. Educational appropriateness, on the other hand, considers the nature of program content--and whether it is "worth knowing" by children (1991, p. 101), whether there are situations where they can use this content, and whether it will help them solve problems in everyday situations. The implementation of any unit of experience requires that the utility of the manipulative activity be assessed in terms of conceptual appropriateness, and not simply hands-on occupation, as the aims of science experiences are planned.

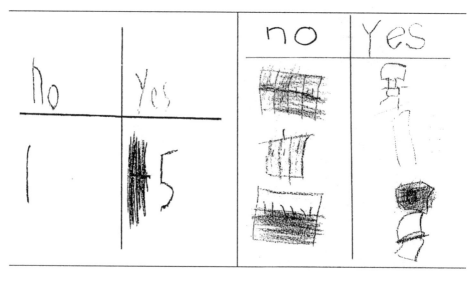

Figure 5. A Kindergarten Science Journal Entry for Recording the Number of Objects Attracted to a Magnet

Figure 6. A Kindergartener's Drawing of Objects Attracted to a Magnet

THE USE OF INFORMATION LITERACY

Is familiarity with the forms and structures of information texts lacking in American language arts programs? Story writing still prevails as the perennial writing assignment "...at precisely the time when the ability to read and write exposition is, arguably, becoming more critical in our society" (Moss, Leone & Dipillo, 1997, p. 419). While children need opportunities to understand and use informational writing, they also need to develop ways of reasoning that helps them investigate the world using the logic and methods of science undergirding such texts. Children's science literacy must go beyond simply telling "stories" and recording results to include explanation and interpretation if children are to engage in and understand science as scientists do.

Through the use of emergent science literacy, young children can be helped to trace scientific uses of logic using their talk, drawing and writing. For example, during a unit about leaves in his kindergarten Sam used talk to explain his understanding of plant growth. He had been wandering around the classroom during circle time as the class examined a leaf and discussed its relationship to a tree, to water, light, and air. Stopping suddenly, Sam said:

> Hey! Maybe this spring we could um um--while it's turning green--then we could bring it to school and THEN we could show the class how it's turning *green*!

Sam hesitated when asked, "Sam, what if we take the leaf off the tree and we bring it in? Would it still work?" Sam responded:

Yes...well, I guess it would NOT work, I just *remembered*, class!

Turning to face the other children, he explained with authority:

Cause the tree would be giving the leaf oxygen...'cause see the tree would be giving the leaf food and then the leaf would just die!

He began again to pace and to make motions with his hands as he spoke:

If we pick the leaf off the tree this spring then it will just *die*. So we have to go out to the tree and we have to *look* at the leaf so--if we're gonna do this *really big* we're gonna have to bring a whole tree to the classroom...and do you think that would be easy? NO!

Sam then began to make a chopping motion. Watching this, his teacher continued, "If you chop the tree down, would it work?" "No!" he said, as if the response were now simple and self-evident. Sam pointed to the chart of a tree by the teacher and explained, "The food comes down here and it goes underground and then it comes up into the tree." He traced the roots and the tree trunk up to the leaves. Here, Sam's technical talk and the teacher's questioning helped him to build understanding by means of two kinds of scientific discourse: an accurate oral explanation of the relationship between the leaf and the tree, and a description of how a process takes place. On his science journal page (shown in Figure 7), Sam communicated his understanding by recharacterizing the elements required for plant growth as super heroes. He drew "air man," (words dictated to the teacher) "light man," (labeled with "L") and "water man" (labeled with "W"). Sam used the page to contextualize his oral explanation, visually depicting an imaginative motif that conveyed the active, living, moving nature of the elements that make plants grow.

Similarly, Nicholas used his science journal to explain the differences between the hard and soft rocks he examined during another science experience; this is shown in Figure 8. He drew a hard and a soft rock, one labeled "D" for diamond and the other "C" for crystal. He also indicated the properties of the diamond: it is "very hard" ("VH") and "shiny" ("S"). He then drew a clenched fist above the diamond to illustrate an attempt to break it. He indicated the result of this process using mirror-writing: "ON" (i.e., "NO"), the diamond will not break if a fist is applied to it. A crystal, shown at the top of the page with its facets, will break if a chisel hits it, as indicated by the dotted line that Nicholas extended from the point of the chisel. Although the children had not attempted to chisel either a diamond or a crystal, Nicholas used drawing and invented spelling to record a process and a result, giving information not empirically gathered.

Martin refers to models of scientific discourse as "world views" (1990, p. 83) through which relationships of parts to wholes are specially articulated through the eyes of the scientific community. This includes the use of technical vocabulary and compressed grammatical structures different from those most common in fiction, for example. Table 6 suggests a compilation (adapted from Martin, 1990; Derewianka, 1990; Foster & Heiting, 1994; Britsch, 1997) of some ways in which scientific discourse may be organized, ways in which children might learn to present information through talk, writing and/or drawing.

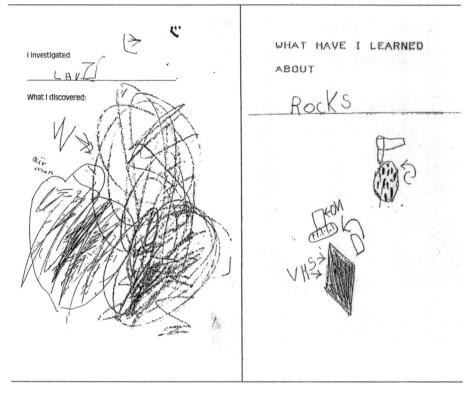

Figure 7. Sam's "Air Man," "Water   Figure 8. Nicholas's Diamond and
Man," and "Light Man"   Crystal Page

While several of these structures presumably reflect cognitive processes, others merely reflect the procedure in which the child engaged. Assessment is limited to noting the ways in which children give information on the page or through oral language. The children's ideas as well as the means through which these are expressed can then be recorded systematically, as illustrated in Table 7.

A CAVEAT

Any program for young children will include more than a single type of assessment in order to characterize conceptual and literate development over time in a variety of situations involving many kinds of tasks. The nature of the teacher's criteria, or foci, is critical and reflects that teacher's own definition of science and of which literacy tools are considered valuable. In other words, our assessments reflect what *counts* in each of our classrooms.

Table 6. Possible Models of Scientific Discourse for Young Children

Model	What the Discourse Does
PROCEDURAL/ HOW TO	Information is presented through a recount of a sequence of actions.
ENCYCLOPEDIC	Information is presented through a restatement of content from an outside source without interpretation or logical reasoning by the author.
EXPERIENTIAL	Information is presented through a description of an investigation carried out by the author.
IMAGINATIVE	Information gained from the science investigation is used to create a world apart from the immediately observable.
INTERPRETIVE	Information is presented by giving or questioning meaning not immediately apparent among sets of empirically gathered data.
LOGICAL	Information is presented inductively or deductively through a description that gives order to data.
EXPLANATORY	Information explains why something is so or how something works.
CLASSIFICATION	Information is classified according to subtypes.
SCIENTIFIC METHOD	Information demonstrates not only the steps taken in scientific inquiry but also: • recognizes a problem • details observations • recognizes patterns • generalizes • formulates hypotheses • explains why a result has occurred

In some classrooms, for instance, the children's questioning might be highly valued, or their ability to recognize a problem instead of merely having the object of study placed before them with prescribed questions to answer and tasks to complete. As teachers, we might ask ourselves how we *expect* children to explore and how we *want* them to present information. We need to be aware of the extent to which our assessment criteria reflect our own values or private hopes about what constitutes learning, emphasizing what *we* want children to accomplish instead of extending their own resources. Are we satisfied with science journals that present information operationally by simply recounting a sequence of actions? Does a restatement of content from a textbook, encyclopedia, or other outside source suffice for us as "research"? Conversely, do we foreground "logical thinking" in assessment? Do we push too many discussions or activities into the same inductive or deductive mold because the ability to order data has greater cachet?

Table 7. Sample Individual Assessments of Science Discourse Functions

Child's Name	Function	Channel	Ideas
Sam	*Explanatory*	*Oral language Drawing, Dictation, Invented spelling (word-initial consonants)*	*Water, light and air are needed for plant growth*
Nicholas	*Procedural*	*Drawing Invented spelling (word-initial sound)*	*Properties of hard rocks versus soft rocks contribute to results of process of breakage.*

Just as Dyson warned against the "reduction of curricula to activities" (1986, p. 136), the assessment of good science teaching and learning should not be confined to the use of checklists. Teaching is, after all, not a set of procedures but instead a set of relationships. There are, as Bussis (cited in Dyson, 1986, p. 142) points out, "...neither prescriptions for action nor checklists for observation to assure intelligent and responsive teaching." Instead, intelligent assessment that is respectful of children's capacities must be guided by teachers' own knowledge of what they are trying to teach and by an understanding of both their own and the children's human resources (Dyson, 1986).

Young children can engage in the scientific process. They are active thinkers who constantly relate the elements of their varied experiences. As teachers, our assessment need not push children to simply find facts to fit well-established contexts. Instead, children need opportunities to relate their manipulative experiences with science phenomena to other, known contexts and concepts. This couches scientific and literate processes in activity that is useful and familiar instead of presenting them as isolated activities that mean little to the children because they are not the result of the children's own activity (Piaget, cited in Duckworth, 1964). The value of science literacy for young children lies in helping them to pull science into themselves. Effective assessment then couches the child's literacy engagement in the context of a complete narrative of the individual's ongoing learning, a narrative with a beginning that precedes the current activity and an end that will follow it. Respectful assessment views learners "...as bathed in the light of their whole biography" (Holquist, 1990, p. 37). Sam's talked-through explanation, Nicholas's labeled processes, and Deanna's stories each provide part of the plot, giving direction to both child and teacher.

REFERENCES

Bear, D.R., Invernizzi, M., Templeton, S. & Johnston, F. (1996). *Words their way: word study for phonics, vocabulary, and spelling instruction.* Upper Saddle River, NJ: Prentice Hall.

Brewer, J. (1995). *Introduction to early childhood education: preschool through primary grades.* Boston: Allyn & Bacon.

Britsch, S.J. (1997). [E-mail and curricular literacy coding categories]. Unpublished raw data.

Clay, M.M. (1975). *What did I write? Beginning writing behaviour.* Auckland: Heinemann.

Derewianka, B. (1990). *Exploring how texts work.* Rozelle, Australia: Primary English Teaching Association.

Duckworth, E. (1964). Piaget rediscovered: a report of the conference on cognitive studies and curriculum development. In R.E. Ripple & V.N. Rockcastle (Eds.), *Piaget rediscovered.* Washington, D.C.: U.S. Department of Health, Education and Welfare, Office of Child Development.

Dyson, A.H. (1986). Staying free to dance with the children: the dangers of sanctifying activities in the language arts curriculum. *English Education, 18,* 135-145.

Dyson, A.H. (1989). *Multiple worlds of child writers: friends learning to write.* New York: Teachers College Press.

Foster, G.W. & Heiting, W.A. (1994). Embedded assessment. *Science and Children, 32,* 30-33.

Gatzke, M. (1991). Creating meaningful kindergarten programs. In B. Spodek (Ed.), *Educationally appropriate kindergarten practices* (pp. 97-109). Washington, D.C.: NEA Professional Library.

Graves, D.H. (1983). *Writing: teachers and children at work.* Portsmouth, NH: Heinemann.

Halliday, M.A.K. (1980). Three aspects of children's language development: learning language, learning through language, learning about language. In Y.M. Goodman, M.M. Haussler & D.S. Strickland (Eds.), *Oral and written language development research: impact on the schools* (pp. 7-19). Urbana, IL: NCTE.

Holquist, M. (1990). *Dialogism: Bakhtin and his world.* London: Routledge.

Martin, J.R. (1990). Literacy in science: learning to handle text as technology. In F. Christie (Ed.), *Literacy for a changing world* (pp. 79-117). Victoria, Australia: Australian Council for Educational Research, Ltd.

Morrow, L.M. (1997). *Literacy development in the early years: helping children read and write* (3rd edition). Boston: Allyn & Bacon.

Moss, B., Leone, S. & Dipillo, M.L. (1997). Exploring the literature of fact: linking reading and writing through information trade books. *Language Arts, 74,* 418-429.

Schickendanz, J.A. (1986). *More than the ABCs: the early stages of reading and writing.* Washington, D.C.: NAEYC.

Tough, J. (1979). *Talk for teaching and learning* (1985 imprint). London: Ward Lock Educational.

DANIEL P. SHEPARDSON AND SUSAN J. BRITSCH

8. TOOLS FOR ASSESSING AND TEACHING SCIENCE IN ELEMENTARY AND MIDDLE SCHOOL

In this chapter we first provide an overview of assessment in elementary and middle school classrooms: we describe the purposes for assessment, outline a strategy for planning assessment in science, discuss the difference between evaluation and grading, and propose a view of assessment as profiling. We next present an overview of different types of assessments, ranging from open-ended response tasks to practical tasks. We then discuss specific methods and formats for assessing children based on their science activity and the products they produce within those activities, including the assessment of science attitude, and peer and self-assessment. We conclude the chapter by talking about children's self-produced journals as a teaching, learning, and assessing tool in science.

INTRODUCTION TO SCIENCE ASSESSMENT IN ELEMENTARY AND MIDDLE SCHOOL CLASSROOMS

Teaching and assessment are closely intertwined in classrooms. Although the purpose of teaching is to help children learn, assessment has multiple purposes:

- Monitoring children's progress
- Diagnosing children's understandings, abilities, and difficulties
- Providing evidence for evaluating and grading children
- Evaluating teaching and programs
- Reporting to parents and others on children's performance and progress
- Providing children with feedback
- Informing pedagogy
- Ensuring accountability

Assessments may be diagnostic, formative, or summative. Assessment for diagnostic purposes determines children's performance abilities in a particular science domain, such as conceptual understanding or inquiry capabilities. Diagnostic assessment uses the information on children's performance to guide pedagogy. Diagnostic assessment may also be used in conjunction with summative assessment to determine change and progress in children's performances. Formative

119

Daniel P. Shepardson (ed.), Assessment in Science, 119—147

assessments are administered on a regular basis throughout a unit of instruction and provide the teacher with feedback on the teaching and learning process. Pedagogically, formative assessments assist teachers in adjusting instruction to align with children's abilities, understandings, and skills. Formative assessments may be used to guide instruction (non-graded) and/or evaluate children's performances. Summative assessment summarizes children's performances. Summative assessments may be administered at the end of the unit as a culminating assessment, or may reflect the accumulation and summary of formative assessments reported in the form of a profile record. Diagnostic, formative, or summative assessments may be formal or informal. Formal assessment involves the recording of assessment results and is preplanned, while informal assessment results tend not to be recorded and are often not planned.

EVALUATION AND GRADING

Although assessment provides information about children's performance, evaluation is the process of determining what the information means. Evaluation involves making judgements about a child's performance based on criteria or standards (criterion-referenced) or comparison to other children's performances (norm-referenced). Evaluation is the process of drawing conclusions about student performance based on the assessment data collected, grading is the assignment of a letter, numerical score, or percentage to the student or the student's work. An important question about grading, often raised by teachers, is: what proportion or percentage should the different assessment tasks contribute to determining a student's grade? In essence, should the assessment tasks be weighted? If so, how? Should assessment tasks be weighted by the type of student performance assessed? Should assessment tasks be weighted by the type of assessment? How does the content of study influence grading? How do the domains of science learning influence grading? These questions can only be answered based on what is valued in the science classroom; for example, if the assessment task is used to organize the grading scheme then the following would be one such scheme:

Open-ended tasks	15%
Drawing and writing tasks	30%
Assignments and projects	15%
Practical tasks	25%
Traditional quizzes and examinations	15%

This grading scheme places less emphasis on assignments and projects and greater emphasis on drawing and writing tasks; it values practical tasks more than quizzes or examinations. The grading scheme could also be structured around the domains of science:

Conceptual understanding	30%
Factual knowledge	20%

Inquiry and science processes	30%
Science attitude	10%
Cooperative and personnel skills	10%

The weighting scheme reflects the teacher's belief about the most import way of assessing student performance and the teacher's values for science learning.

PLANNING FOR CLASSROOM ASSESSMENT

Although assessment is a component of classroom activity, it is often not planned as an aspect of the classroom, but it is thought of as a happening, an event separate from instruction (Harlen, 1988). Assessment, however, needs to be viewed as an integral component of the classroom environment, a system aligned with and integrated into instruction. The National Science Education Standards, Teaching Standard C notes that, "Assessment tasks are not afterthoughts to instructional planning but are built into the design of teaching" (National Research Council [NRC], 1996, p. 38). Assessment, like instruction, requires planning to create a system that articulates the strategy for assessing children within the context of instruction. The assessment strategy should outline:

- The purposes for assessment
- The instructional goals and objectives that will be assessed
- The mix of assessments that will be used (the format and method of assessment)
- The points within the curricular/instructional units when the assessments will be administered
- The science domains that will be assessed
- The criteria or standards that will be used to evaluate student performance
- The contribution of these assessments to the grading process

The assessment grid presented in Figure 1 may assist teachers and administrators in thinking about and planning an assessment system for a classroom or school. The assessment matrix illustrates the relationship among the science domains (what to assess), instruction and curriculum (when to assess), and assessment format and method (how to assess). Powerful classroom assessment practice requires that multiple assessment sources be used, that multiple student traits or science domains be assessed, that assessments be standards-based, and that instructional goals and pedagogy be aligned with assessment practice. Further, the purpose for assessment must be clearly articulated. Deciding what to assess is determined in part by the instructional and curricular goals and objectives of the school or classroom; however, the main domains of science learning include: conceptual and factual knowledge, inquiry and science process skills, science attitude, and cooperative and personnel abilities. State and national standards should be used as tools for identifying the criteria that students should meet within these domains.

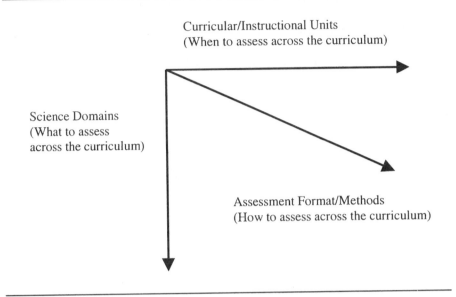

Curricular/Instructional Units
(When to assess across the curriculum)

Science Domains
(What to assess
across the curriculum)

Assessment Format/Methods
(How to assess across the curriculum)

Figure 1. Assessment System Planning Grid

ASSESSMENT AS PROFILING

The concept of profiling presented here defines assessment as a process of creating a picture of student performance or ability over time. It involves thinking about assessment as a system consisting of a variety of assessment tasks and formats used to construct a picture of student ability and progress based upon a variety of student traits or science domains. This is called profiling. Profiling is not an assessment task, but the means of organizing and reporting student performance. The profile reports on similar student traits or science domains across the curriculum, illustrating students' performance over the academic year. It is a means for going beyond the traditional grade book, yet complementing the grade book. Assessment as profiling is in alignment with the National Science Education Standards, Teaching Standard E (NRC, 1996), calling for teachers to report student achievement in ways that go beyond grades. Drawing from the work in Britain (Balogh, 1982; ILEA, 1984), a good profile in science would:

- Record the assessment of the following domains of science: science process and inquiry skills, factual and conceptual understanding, problem-solving ability, cooperative learning ability and skills, and science attitude.
- Record the assessment of the following personnel domains: motivation and commitment, responsibility, perseverance, and self-confidence.

- Report students progress on the science and personnel domains over time from the beginning to the end of the academic year. It is not necessary that every assessment task measure each domain, but that over time each domain is assessed several times and recorded on the profile.
- Present a structured form and contain the same information for each student.
- Be made public to both students and parents.

Assessment as profiling requires one to think differently about what is assessed: what do we want students to know and be able to do in science? Profiling requires one to think about what domains in science are valued and how those domains are assessed across the science curriculum. An assessment profile assists in organizing and focusing the classroom assessment process; it builds an assessment system that connects assessment tasks across the curriculum based on key domains of science learning versus an isolated assessment process of individual and content dependent assessments. The intent of assessment as profiling is not to assess students on every science domain every time, as this would be an overwhelming task, but to sample students performance throughout the academic year.

Science assessment as profiling would provide students, parents, and others with more detailed information about student performance and progress than traditional grades or state mandated tests. These profiles would complement the traditional grade reporting system by providing a more detailed basis for the grades students receive. The profiling process, however, does require more teacher time and commitment, and requires an assessment plan that articulates and coordinates assessment with instruction across the curriculum.

TYPES OF ASSESSMENTS

The types of science assessments used in elementary and middle school classrooms may be grouped into the following categories: practical tasks, open response tasks, selected response tasks, constructed response tasks, product tasks, and process tasks. We describe below two of these alternative assessment types: practical tasks and open response tasks. Following these descriptions, we share more detailed examples of assessment methods and formats based upon our collaborative work with elementary and middle school teachers of science.

Practical Tasks

Practical tasks are specific tasks designed to engage children in manipulating materials, using equipment, and observing phenomena from which their performance abilities are assessed. Practical tasks tend to be administered as summative assessments for the purpose of assessing children's conceptual understanding and/or their inquiry ability or process skills. Summative assessments, however, can also serve as instructional activities embedded within instruction, or as stand-alone formative assessments. Practical tasks involve children in conducting investigations, planning investigations, or both. Children's work in practical tasks may be assessed through the use of observational checklists, interviews, or through the analysis of children's products.

Open Response Tasks

Open response tasks are paper-pencil tasks that provide children with information and ask children to apply their understandings to the situation described in the task. They do not ask children to re-state facts; they allow children to respond in a variety of ways to explain and apply their ideas. Open response tasks have many formats ranging from draw and explain tasks to interpretation of data and graphs, to open-response writing tasks. Although open response tasks are often administered as summative and sometimes as formative assessments, they need not be used as assessments at all. Open response tasks could easily serve as instructional activities, providing a basis for children's written and oral discussions.

SCIENCE ASSESSMENT METHODS AND FORMATS

In this section we describe in detail different methods and formats for assessing children's science performance. We first discuss issues involved in peer and self-assessment, and assessing science attitudes. The remaining assessment methods and formats emphasize children's science activity and the products they produce within that activity. We specifically look at assessing children's science learning through their graphic products (writing and drawing), their talk (oral interviews), observation of their activity within science investigations, and their planning and conducting of science investigations. The last section of the chapter outlines the use of children's self-produced journals as tools for teaching, learning, and assessing.

Peer and Self-Assessment

Self-assessment is designed to engage children in reflecting upon their own performance (their collaborative and individual work) to promote responsibility for learning, and to build self-confidence. Engaging children in self-assessment has several benefits: it provides a means to evaluate the match between instruction and children's ability; it motivates children to perform better, and it makes explicit the purpose for the activity (Broadfoot & Towler, 1992; Ollerenshaw and Ritchie, 1997). Children can and should be involved in assessing their own science learning, however, teachers must assist children in developing their skills and abilities to engage in meaningful self-assessment. It is essential that children be given the language and the tools that will help them be successful in reflecting on their science learning. In addition, clearly stated criteria must be provided to support children's self-assessment; such criteria may be constructed by the teacher or in cooperation with the children. Although children need not self-assess for every science activity, it is imperative that self-assessment occur on a regular basis in order to develop children's abilities to self-assess and to communicate the importance or value of self-assessment.

The incorporation of peer assessment aligns the assessment process with instruction. Children are often required to collaborate on the completion of science activities and thus should have opportunities to conduct peer assessments. Peer assessments may be used to provide feedback about group work and the quality of academic work, and may establish a level of individual accountability. Further, peer and self-assessment can inform pedagogy and provide teachers with insight into children's understandings and abilities. The *National Science Education Standards* (NRC, 1996) paint a picture of peer and self-assessment in which the teacher assists children in assessing and reflecting on their own and their peers' science learning by formulating strategies and tasks that guide children in these processes.

In the example provided (Table 1) the child would be required to describe in writing how they demonstrated the performance during the investigation. The teacher and child then meet to discuss the rating, with the teacher approving the rating or negotiating a change. The observational checklist (see Table 8) may be reformatted into a self-assessment where children assess their performance during an investigation. As with the teacher's observational checklist, it is not essential that children assess themselves on each performance category for every investigation but that, throughout the academic year, children assess themselves on each performance category several times. Children could also complete the self-assessment checklist as a form of peer assessment.

Table 1. Self-Assessment Record for Conducting an Investigation

Performance Categories	Demonstration	Self-Rating	Teacher Approval
Controls variables and measures appropriate variables			
Uses appropriate degree of precision in measuring			
Makes notes or records observations during investigation			
Makes accurate observations			
Appropriately organizes data and information (data table)			
Makes appropriate graph, chart, or drawing			
Follows investigation plan or procedures			
Obtains information from existing resources			
Follows safety procedures			

Rating scheme: S = satisfactory, I = improvement needed, N/A = not applicable

Assessing Science Attitude

Ollerenshaw and Ritchie (1997) point out that the assessment of science attitude needs to consider qualities that are desirable when conducting scientific work. Important aspects of working scientifically include: curiosity, respect for evidence, willingness to change ideas based on evidence, perseverance, critical reflection, willingness to collaborate, and meticulousness. The *Benchmarks for Science Literacy* (American Association for the Advancement of Science [AAAS], 1993) also indicate the importance of developing children's attitudes towards science and scientific work:

> By fostering student curiosity about scientific, mathematical, and technological phenomena, teachers can reinforce the trait of curiosity generally and show that there are ways to go about finding answers to questions about how the world works. . . . Balancing open-mindedness with skepticism may be difficult for students. . . . Students hearing an explanation of how something works proposed by another student or by teachers and other authorities should learn that one can admire a proposal but remain skeptical until good evidence is offered for it (AAAS, 1993, p. 284).

It is important to understand that attitudes are contextual; that is, children's attitudes towards science and scientific work may change based on the science topic under study or the nature of the science activity. Although attitude is given importance as a goal of science education, it is rarely assessed in classrooms. Unfortunately, children's science attitudes cannot be improved without an understanding of existing attitudes and certainly cannot be given status in classrooms if it is not assessed. The dilemma many teachers face with the assessment of science attitudes relates to grading; however, science attitudes could be assessed for non-grading purposes. Harlen (1993) suggests indicators of science attitude that could be utilized in assessing children's science attitudes through observational checklists, child interviews, and peer and self-assessment:

Curiosity
Noticing and attending to new things and situations
Showing interest through careful observation of details
Asking questions
Using resources to find out about new or unusual situations

Respect for Evidence
Checking evidence that does not appear to fit the pattern of other findings
Querying a conclusion or interpretation where there is insufficient evidence
Treating ideas as provisional and being open to challenge by further evidence

Willingness to Change Ideas
Changing an existing idea when there is convincing evidence
Considering alternative ideas to their own
Seeking alternative ideas versus accepting the first idea
Realizing that it may be necessary to change an existing idea

Critical Reflection
Willingness to review what they have done and consider improvements
Consideration of alternative procedures
Identification of positive and negative aspects of their investigation
Use of critical reflection of previous investigations to plan and conduct future
    investigations

Assessing Children's Graphic Products

Children's graphic products (writing and drawing) in science can provide teachers with powerful insight into children's understandings and thinking or reasoning processes. Like a mirror, they reflect children's ideas and ways of thinking, and serve as a guide to children's understandings (Elstgeest, Harlen, & Symington, 1985). Children's graphic products not only provide evidence of performance, but also serve the following pedagogical purposes:

- Provide a purpose and form for student work
- Encourage active learning
- Allow children the opportunity to communicate their ideas and thinking
- Reflect children's understandings and thinking
- Promote children's mental activity
- Provide a means for recording data and information
- Mediate children's knowledge construction
- Improve children's observations
- Value children's ideas and thinking
- Provide children with a sense of ownership (Elstgeest, Harlen, & Symington, 1985)

Younger children's graphic products contain unconventional symbols that only they may understand. These products may contain imaginative additions and embellishments that couch the science phenomena in a fictive context (for more detail on assessing young children's graphic products see the chapter "Assessment for Emergent Science Literacy in Classrooms for Young Children" by Britsch). The drawing and writing of some young children may represent their experiences with or preconceptions of the phenomenon. Other children's writing and drawing may be contextualized by the classroom science investigation. Different children use different world views to make meaning about school science activities (Britsch & Shepardson, 1996). For example, first-grade children investigating the life cycle of beetles and butterflies drew pictures of the mealworms and beetles and caterpillars as observed in the containers; however, as the butterflies emerged from their chrysalis, some children reconceptualized the investigation in terms of their experiences with caterpillars, cocoons/chrysalis, and butterflies/moths (Shepardson, 1997). Sally's drawing depicting a flower and a plant with a cocoon, illustrates her reconceptualization.

Teachers need to assist children to attend to the important features of the phenomenon based on the purpose of the activity. If instruction requires children to produce written products and to record detailed and accurate observations, Elstgeest, Harlen, and Symington (1985, p. 111) suggest the following to assist children in their graphic productions:

- Talk about the purpose of the drawing or writing activity.
- Talk about what features or details children should observe or attend to.
- Talk to children about their drawings and writing, asking questions that guide their activity and observations of important and relevant details.
- Have children explain their drawings to uncover understandings and ways of thinking not represented in the child's graphic product.
- If the purpose of the drawing is to attend to and record detailed observations, talk about drawing observations versus drawing preconceptions.

If the purpose for drawing is to have children place their understandings into a context for meaning, contextualizing their understandings of the science activity, then children should be encouraged to make drawings that reflect their experiences of "real world" contexts (Shepardson, 1997).

There is also a delicate balance between having children produce authentic products that represent their activity, understandings, and thinking processes and products that reflect closed-ended writing or drawing exercises that emphasize an answer to a teachers' question. This is not to say that closed-ended writing and drawing cannot sometimes be a valuable teaching-learning-assessing activity, but that the purpose of the task and the outcomes differs from that of more authentic writing-drawing activities. For example, Elstgeest, Harlen, and Symington (1985) described a situation where a teacher asked students to draw a picture of what a balance looks like if one side is heavier. Children interpreted the activity as a test and did not use actual balances to answer the question. For some children, the drawing represented an incomplete picture of their understanding of balances. As Elstgeest, Harlen, and Symington (1985) noted, a more effective question for structuring the children's activity and written products would have been to ask children to write and draw about their own experiment, what they did, what they found out, and what they learned. This approach to children's activity and production of graphic products embeds assessment in instruction as a single pedagogical-assessment activity.

Older children's graphic products produced during science activities may also be assessed from the perspective of both writing and drawing so that the assessment process becomes embedded in the context of the instructional activity. Although numerous performance categories could be identified for assessing older children's products, Table 2 illustrates five drawing and five writing categories that could be used within a science activity. The assessment matrix for children's drawing and writing illustrates the potential for assessment based on graphic products, and is by no means the only method to be used in assessing children's science learning.

Children's talk about their work and products is also essential to understanding children's science learning and progress and should be considered in the assessment process.

Table 2. Assessment Matrix for Children's Drawing and Writing

Performance Categories	Criteria	Rating Scale/Scoring System
Drawing 1. Sequence of activity		
2. Sense of scale and relationship between objects		
3. Level of detail		
4. Relationship of drawing to writing		
5. Carefulness		
Writing 1. Level of description/detail		
2. Use of words and language		
3. Vocabulary		
4. Factual understanding		
5. Conceptual understanding		

In the balance example, the child has made four drawings that illustrate the sequence of his investigation. He has also written two brief statements that describe his findings, "If things weight the same they will balance! If one is heavier it will go down like tetertator (sic)." The assessment of the child's product is shown in Table 3. Perhaps the most important aspect is the insight it provides into the child's conceptual understanding. It appears that the child understands that the weight of the objects on each arm determines the balance, but he does not show an understanding that the distance from the fulcrum as well as the weight of the objects determines whether the objects will balance. This result has several implications. Does the child not conceptually understand balances or is it simply that he did not investigate the fulcrum-object weight-distance relationship? It would appear that the child did not investigate the relationship between the weight of the objects and the distance from the fulcrum; none of the drawings show a change in the distance from the fulcrum. On this basis the teacher might re-engage the child in the balance activity and focus on the fulcrum-object weight-distance relationship. In the future, more specific directions and expectations for completing the balance activity are probably needed. This situation also illustrates an important point: assessment activities do not always have to result in grades, but can inform pedagogy and diagnose children's conceptual development.

Table 3. Example Assessment of Children's Drawing and Writing

Drawing

1. Sequence of activity	*Shows sequence the objects were tested.*
2. Sense of scale and relationship between objects	*Shows relationship between objects tested (wood block, ball, and spoon) and the change in balance; changing the block (#4) doesn't change the outcome. Objects show scale.*
3. Level of detail	*Show detail of objects and balance.*
4. Relationship between drawing and writing	*Writing relates to drawing.*
5. Carefulness	*Drawings are neat and specific.*

Writing

1. Level of description/detail	*Weight of objects not indicated in #1; others indicate that the block is heavier.*
2. Use of words and language	*Labels drawings, describes relationships between objects: "Weight the same they will balance;" "One heavier it will go down."*
3. Vocabulary	*Heavier, weight, balance, and "tetertator"*
4. Factual understanding	*Understands that spoon and ball weight the same, block is heavier than ball and cup; changing the block (#4) doesn't change the outcome.*
5. Conceptual understanding	*Realizes that the weight of the objects on each arm determine if they balance. Does focus only on weight. Does not show understanding of the relationship between the distance from the fulcrum and weight of an object in determining if the objects will balance.*

Asking older children to write about their science activities can provide insight into their understandings; however, as an assessment activity it requires more structure than simply asking children to write about the science experience. Writing as assessment requires thoughtful guidance to structure the child's writing. For example, in wood car rally where children plan and conduct a computer simulation investigating the effect of the slope of the ramp, the distance up the ramp the car is placed, the type of car lubricant, and the type of car body has on the distance a car travels, questions to structure children's writing might include:

- How did changes in the slope of the ramp affect the distance traveled by the car?
- Describe what you think would happen if the slope of the ramp is increased from 5 to 10 to 15 degrees.
- If you were a race car driver, which lubricant would you use and why?

- If you were to explain to a fourth-grader about how friction affects the distance a car would travel down the ramp, what would you say?

Freedman (1994) describes a process for constructing more extensive open-ended writing questions based on five formats appropriate to science; these are based on critical thinking skills: analysis, comparison, description, evaluation, and problem solving. Freedman (1994) identifies a five step process in writing open-end questions: 1) review the curriculum for concepts or topics that lend themselves to the question formats, 2) choose the question format based on the thinking skill to be assessed, 3) develop the writing prompt that describes the situation, 4) develop the directions for writing, and 5) develop the scoring rubric considering the conceptual understanding, content knowledge, and critical thinking processes. An example of an open-ended writing question follows:

Writing prompt: A housing developer is planning to build a subdivision near the Wabash River, but is concerned that the subdivision might be damaged by erosion. Based on your stream table investigations, the developer has asked you for help in determining what environmental conditions should be measured to determine if erosion would be a problem.

Writing directions: You are to write a report describing the environmental conditions the developer should measure and why. Your report should include results from your stream table investigations and information from books and other resources that support your recommendations.

Although the actual scoring rubric would depend on the children's instructional experiences, a possible top level performance might reflect:

Conceptual understanding: Water from the river and rain will shape and reshape the surrounding earth's surface by eroding soil in some areas and depositing them in other areas.

Content knowledge: Erosion, sediment, deposition, slope, soil, sand, and clay.

Thinking/reasoning processes: Statements supported with evidence from the stream table investigation and facts found in books and other resources; sources are identified.

Assessing the Planning and Conducting of Investigations

Engaging children in science as inquiry requires children to plan and conduct investigations; however, these processes are rarely assessed. Instruction needs to involve children in planning their own investigations and assess these plans. If not, children cannot develop their skills and abilities to plan. Engaging children in planning science investigations also helps them to think through possible actions and alternative approaches to their investigations; it helps them rethink ideas in light of the evidence and the procedures for conducting the investigation, and it helps them link procedures to results (Harlen, 1985).

The assessment records shown in Tables 4 and 5 illustrate two variations for assessing children's performance abilities in planning and conducting investigations. Both are based on performance categories that reflect the key elements of planning and conducting science investigations: the investigation question, understanding the variables involved in the investigation, the procedures (plan) for conducting the investigation, thinking about how the results will answer the research question, how the data is organized, how the data is transformed (usually as a graph), how the data is interpreted, and how the data is explained using scientific concepts and principles.

Table 4. Assessment Record for Planning and Conducting an Investigation: Specific Criteria Form

Assessment Record For:_____   Score: _____
Student:_____

Performance Category	Specific Criteria (How Demonstrated)	Rating[1]	Score[2]
Investigation question			
Understanding variables			
Procedure (Plan)			
How results answer question			
Data organization			
Data transformation			
Data interpretation			
Explanation of data			

[1]Rating scheme: N/A, not applicable; S, satisfactorily demonstrated; U, unsatisfactorily demonstrated
[2]Scoring system: S = 1 point, U = 0 points
Note that the rating scheme and scoring system may be changed to reflect the goals and objectives of instruction.

Table 5. Assessment Record for Planning and Conducting an Investigation:
Generalized Form

Assessment Record For: _____   Score:_____
Student:_____

Performance Category	Scoring[1]	Comments
Investigation question		
Understanding variables		
Procedure (Plan)		
How results answer question		
Data organization		
Data transformation		
Data interpretation		
Explanation of data		

[1]Scoring: N/A, not applicable; 2 (points) no difficulties; 1 (point) some difficulties;
0 (points) unsatisfactory
Note that the rating scheme may be changed to reflect the goals and objectives of instruction.

The first assessment record (Table 4) requires the teacher to identify the criteria specific to the investigation and then rate the children's performance based on these criteria. Because children may be planning and conducting different investigations within a unit of study, the assessment record would vary to reflect the specific criteria of each investigation. The rating is then translated into a score, which can be weighted. The rating scheme and scoring system are established by the teacher prior to the investigation based on the goals and objectives of instruction. In the example provided, students would need to demonstrate all criteria for a performance category to receive a satisfactory rating. Failure to meet all criteria would result in an unsatisfactory rating. Each performance category is equally weighted so that the interpretation and explanation of data would account for 25% of the score; this could place an overemphasis on planning and conducting the investigation.

The second example (Table 5) is perhaps less time intensive, but is also less explicit about the specific assessment criteria used. It does, however, require that explicit comments about students' performance for each category be written to substantiate the student's score. Thus, for any investigation, each assessment record requires that the teacher's written comments be similar to the specific criteria stated prior to the investigation.

In addition to assessing students' performance on individual investigations, it is useful to develop an assessment profile of students progress in planning and conducting investigations. To do this, teachers may wish to develop an assessment profile for each child similar to the profile record illustrated in Table 6. These performance categories should align with those in the assessment record. By creating an assessment profile, teachers have a record of children's progress over time.

Table 6. Assessment Profile for Planning and Conducting an Investigation

Student: _____

Performance Category	Inquiry Activity and Date		
	Wood Car Rally (date)	*Paper Towels (date)*	*Dissolving (date)*
Investigation question			
Understanding variables			
Procedure (Plan)			
How results answer question			
Data organization			
Data transformation			
Data interpretation			
Explanation of data			

Rating scheme: N/A, not applicable; S, satisfactorily demonstrated; U, unsatisfactorily demonstrated
Note that the rating scheme may be changed to reflect the goals and objectives of instruction.

Children's planning and conducting of investigations can also be assessed based on the products they produce within the activity. In the wood car rally example, children's planning sheets, investigation sheets and posters could be assessed. The planning sheet records the ability to state a research question, identify variables, and develop procedures for conducting the investigation. The investigation sheet provides evidence of children's ability to record data, transform data, interpret data, and explain the results of the investigation. The poster presentation also assess the ability to plan and conduct investigations, conceptual understanding of the investigation, and application of scientific concepts.

In the wood car rally example, children were required to present their research questions, the variables involved in their investigation, a graph of their results, an interpretation of their data, and an explanation based on scientific concepts. An example scoring rubric for assessing an investigation poster is found in Table 7. Based on the poster, the following conclusions can be drawn about the group's performance:

- The research question clearly states the variables involved in the investigation, Indy car, lubricant, and distance traveled from end of ramp.
- The independent (no lubricant, oil, graphite) and dependent (distance traveled) variables are identified, as well as the variables being controlled (slope of ramp, distance up ramp, type and weight of car).
- The use of a bar graph is appropriate for the data collected and the axes are appropriately labeled.
- The graph lacks a title and the vertical axis is not to scale.
- The interpretation reflects the data collected.

Table 7. Example Assessment Rubric for Investigation Poster

Performance Level	Performance Category			
	Research question	Identification of variables	Data transformation	Data interpretation
3	The question is researchable and clearly states the variables involved in the investigation	Identifies all independent, dependent, and controlled variables	Uses appropriate graph for data collected; all axes are labeled and to scale; title and key provided	Interpretation is supported by the data collected
2	The question is researchable but does not clearly state the variables involved in the investigation	Identifies most but not all independent, dependent, and controlled variables	Uses appropriate graph for data collected; than some difficulties in axis labeling and/or scale and/or, title; key not provided	Interpretation is partially supported by the data collected
1	The question is not researchable	Fails to identify all independent, dependent, and controlled variables	Inappropriate graph for data collected	Interpretation is not supported by the data collected

Note: The data explanation column and not attempted row have been deleted because of space constraints.

Although the explanation refers to friction and momentum, it is only partially accurate or complete. There is some confusion in that the lubricant is believed to reduce friction between the wheel and the ramp and it is not clear how this affects momentum. The explanation does not utilize the concepts of potential and kinetic energy.

Based on the scoring rubric in Table 7, this group's poster would receive the following performance level rankings: research question, level 3; identification of variables, level 3; data transformations, level 2; data interpretation, level 3; and data explanation, level 2. These performance level rankings may be converted to a score by treating each performance category the same; adding the performance level rankings derives the group's score (13 out of 15). If greater value is placed on data interpretation and explanation, a system could weight data interpretation and explanation performance level ratings by a factor of two. The group's poster score would then be 18 out of 21.

Observational Checklists and Child Interviews

Teachers may assess children's performances during an investigation through direct observation, checklists to record the observations, and through child interviews. Teachers have informed us that the key to the successful use of observational checklists and child interviews is to limit the performance categories and the number of interview questions to three or four. Which child actions should be observed during an investigation? Drawing on the work of Harlen (1983; 1988) and Hodson and Brewster (1985), indicators of children who plan and conduct investigations successfully:

- Control variables and measures appropriate variable (dependent variable) while conducting the investigation.
- Use appropriate degree of precision in measuring.
- Make notes or records observations during investigation.
- Make accurate observations.
- Appropriately organizes data and information in tabular form.
- Make appropriate graph, chart, or drawing.
- Follow laboratory safety procedures.
- Obtain information from existing resources.
- Follow investigation plan or procedures.

Using these indicators, an observational checklist may be constructed for assessing children's performances during an investigation. The checklist may be used as a profile (Table 8) to illustrate progression in performance over the academic year. Although it is not essential to assess children on each performance category for every investigation, it is necessary to assess each performance category several times throughout the academic year. If a score is desired, the rating scheme may be converted to a score (e.g., a satisfactory rating could receive a score of two and an improvement rating could receive a score of one). Observational checklists reflect children's inquiry and process skill performances and do not provide insight into children's conceptual understandings. Checklists should be used in conjunction with other measures of children's conceptual understandings.

In developing interview questions to ask children during or about the investigation, consider questions that require children to make predictions based on evidence, explain their results, and support their interpretations based on their results. Further, interview questions should ask children to suggest modifications and improvements to their investigation. For example, in the wood car rally activity a teacher might ask: "How far do you think the car will travel if you increase the slope of the ramp from 10 to 15 to 20 degrees? How does the slope of the ramp affect the distance the car travels? Which lubricant is the best? How would you explain your results based on the scientific concept of friction?" Indicators of children's successful responses to these questions should reflect:

- Predictions supported by past evidence or evidence from the investigation.
- Identification of patterns or trends in observations or results.
- Interpretations based on information or data collected.
- Explanations of results using scientific concepts and principles (Harlen, 1983; Hodson & Brewster, 1985).

Table 8. Observational Checklist and Profile for Conducting an Investigation

Student:

Performance Categories	Science Investigation and Date		
Controls variables and measures appropriate variable			
Use appropriate degree of precision in measuring			
Makes notes or records observations during investigation			
Makes accurate observations			
Appropriately organizes data and information (data table)			
Makes appropriate graph, chart, or drawing			
Follows investigation plan or procedures			
Obtains information from existing resources			
Follows laboratory safety procedures			

Rating scheme: + = satisfactorily performed, - = improvement needed, N/A = not applicable

These questions are different from those designed to probe children for their ideas or understandings. If the purpose of the interview is to elucidate children's understandings, then questions need to address the concept of study. Bell, Osborne, and Tasker (1985) have presented extensive guidelines for interviewing children about their understandings. An important element here is to emphasize the child's meaning by using question stems such as: "Why do you think," "Can you explain to me the way you think?" and, "What do you think happens?" For example, a teacher interested in children's understandings about cars and ramps might first ask, "Why do you think the car travels a greater distance as the slope of the ramp is increased?" Depending on the child's response, the teacher might then ask, "What do you think happens to the car if the slope of the ramp is decreased?" The child's responses can then be recorded as evidence of conceptual understanding.

JOURNALS AS TOOLS FOR TEACHING AND ASSESSING

Incorporating children's journals into science teaching can expand our teaching and assessing methods. The use of journals in elementary classrooms, however, has been mostly confined to language arts activities and is rarely extended to science teaching (Elstgeest, Harlen, & Symington, 1985), in part because of an uncertainty about how best to use science journals in the teaching-learning process. Science journals provide an opportunity to access and assess changes in children's understandings and thinking, and obtain a more complete picture of children's understandings of science phenomena (Dana, Lorsbach, Hook, & Briscoe, 1991). To do this we need to examine children's drawing and writing products as they construct and represent their understandings (Doris, 1991) in science journals. But what is an effective way to use children's journals in science teaching? How can we assess children's journals for science learning; in other words, what are the characteristics, or indicators, we look for in children's journals to assess science learning? We offer answers to these teaching-assessment questions drawn from our collaborative research program with teachers and children.

Teaching Science Using Children's Journals

Although children's self-produced journals can be used in science teaching in many ways, we focus here on using children's journals during the investigation of scientific phenomena. First, the journal can be used primarily as a log in which children describe their experimental procedures, record their observations, and report their conclusions. While this approach encourages children to attend to procedural details and to make accurate observations, it also constrains children's focus to the science experience at hand. This final product provides little opportunity for children to elaborate their writing and thinking. This approach provides limited information about children's science understandings and learning, and tends to be confined to the procedural and factual knowledge children draw from the science investigation itself.

Table 9. Instructional Outline for Using Children's Science Journals

Instructional Phase	Children's Activity (Writing and Drawing in Journal)
Pre-Investigation	Explains existing ideas and understandings Describes purpose of the investigation States the question(s) to be answered by the investigation. Makes prediction based upon existing ideas Describes investigation plan/explains procedure
Investigation	Records observations (qualitative, quantitative) Records ideas and thoughts about investigation based on observations Reflects on existing ideas and predictions in light of observations and on going findings Creates drawings, charts, or tables for organizing data
Post-Investigation	Answers question(s) using observations and data collected during the investigation as evidence Uses information and other resources to explain the results or relate to findings Creates charts, graph; transforming the data Reflects on existing ideas and predictions in light of findings and explanations Identifies ways of conducting the investigation differently or improving the investigation Proposes new questions for investigation
Communication	Uses information written and drawn in journal to communicate or share the investigation with others; applies findings to the everyday; may include: ♦ Science conference or convention ♦ Science article ♦ Science book ♦ Science poster ♦ Poetry, songs, or stories

The second approach provides a structure for guiding children's writing and thinking. On this view, the children's science journal is not the end product, but a resource for the creation of a final product. This approach moves assessment beyond procedural and factual knowledge to conceptual understandings and attitudes. The approach is simple and easily implemented, involving four phases of journal activity: pre-investigation, investigation, post-investigation, and communication (see Table 9 for an instructional outline). This approach, unlike the first, requires that more time be devoted to children's journals as a central component in learning science.

Pre-Investigation Phase

The pre-investigation phase lays the groundwork for the investigation and for children's science learning. During the pre-investigation phase, children write and/or draw in their journals to clarify and express their existing ideas about a particular phenomenon. Children then state the purpose of the investigation, present the question(s) to be answered, make predictions based upon their prior ideas and understandings, and plan the investigation or explain its procedures. Pre-investigation journal writing and drawing thus provides an opportunity for children to think about the investigation and make explicit their ideas and understandings for later reflection.

Giving children the opportunity to write about what they already know enables them to solidify their understandings, and provides a base from which the investigation can confirm or challenge their existing understandings. Recording the purpose and stating the questions of the investigation helps focus children's planning and observation. Making predictions helps children examine their own ideas in light of the investigation's questions and findings. Throughout, writing and drawing about the procedure keeps the investigation on track, and helps children link procedural knowledge with conceptual knowledge.

Investigation Phase

The investigation phase involves more than the simple collection and recording of data or observations; it is the backbone of the learning process. It is a time when children organize their data into charts and tables. It is a time when children can also write about their new ideas concerning the investigation and the findings to date. It is a time when children may reflect, in writing, about their initial ideas, understandings and predictions, as well as their investigation procedures. It is a time for externalizing thought through writing and drawing on the journal page. If appropriate, it is a time when children can seek and record information from other sources to help them understand and explain the investigation.

Post-Investigation Phase

The purpose for post-investigation journal writing and drawing is to assist children in interpreting and explaining their investigation results, while reflecting on their existing ideas and understandings. Children can now refer to their prior journal writing and drawing to make sense of their ideas about the investigation. This is the most difficult task for children because it pushes them to draw conclusions based on their own interpretations, using the evidence collected during the investigation. To do this, children must first transform their data into graphs or charts and then use the transformed data to support their written and drawn interpretations. At this time children may use other resources, such as science textbooks, magazine articles, trade books to assist them in composing written explanations of their investigation results.

Finally, children should write reflections about their initial predictions, summarizing what they learned, how their ideas changed, and how they think the investigation relates to the everyday world. They should also reflect on the investigation, identifying ways they could change or improve it; they may also propose new questions for subsequent investigation. This is essential to promoting children's explorations of their own thinking and learning processes, but is often omitted if science journals are used primarily as procedural and observational logs.

Communication Phase

In science, journals and logs are often maintained as a means for recording data and information; but they are seldom used to communicate and share information with others. The science journal typically functions as a tool scientists draw on in order to create other products that communicate their findings to other members of the community. Children's science journals can also function as tools for communication instead of final products in and of themselves. By constructing a final product for communication, children synthesize and apply their new understandings to a new context as they explore the genre of scientific narrative. The nature of the final product depends on the extent of the investigation and on the audience. For example, children may construct a product that is used at a peer "science conference," where they orally present their findings to other child scientists. Alternatively, children could write an article for publication in the classroom or school science magazine; they might develop a science book for other children to use; they might create a science poster, or a poem, song, or story to communicate their findings.

Assessing Children's Science Learning through Journals

New ways of teaching science require assessment approaches that are both seamless with instruction and pedagogical in nature; that is, the assessment processes must function as teaching-learning activities. Assessment must also be multi-trait, assessing more than one domain of children's science learning. The use of children's journals allows teachers to assess children as they engage in activity and learn science, and to assess the multiple dimensions of science learning. Children's science journals enable teachers to assess the domains of conceptual understanding, factual and procedural knowledge, science processes, and attitudes. We first present sample indicators that may be used to assess the domains of science learning. We then present a process for constructing scoring rubrics that assess these science learning domains, followed by suggestions for successful implementation.

Assessing for Conceptual and Factual Understanding

Assessing children's journals for conceptual understanding is not accomplished by looking at single sentences or words, but by analyzing the entire journal over time. This enables teachers to determine children's prior understandings and follow the changes in these understandings throughout the science experience. Children's conceptual understandings are also displayed in their use of written narrative to interpret or explain, while factual understanding may be expressed through labels and descriptions. To assess children's factual understanding, the journal page may also be analyzed for specific content information that clarifies, or unpacks, the meaning of specialized scientific vocabulary.

Assessing for Procedural Knowledge and Science Processes

When assessing children's journals for the use of the scientific process, teachers can analyze the children's writing and drawing for several items:

- The investigative question, predictions, observations (see below for more detail).
- Descriptions of procedures and explanations of material and equipment use.
- The identification and control of variables.
- Representation and organization of data.
- Conclusions based on evidence.

The checklist displayed in Table 10 provides a simple means to assess children's journals for the science processes. It is not necessary to assess every item on the checklist every time, but over time each item will be assessed a number of times.

Although children's observations are often downplayed in the science teaching process, they are essential to the learning of science. It is from the child's observation of science phenomena that the child begins to construct scientific understandings. Children may, however, have difficulties in observing similarities, in easily observing differences, and they may erroneously record deductions as observations. To assess the quality of children's observations, teachers can analyze their journals for the following indicators:

- Actual observations versus deductions.
- Level of detail (qualitative and quantitative).
- Notations of similarities and differences.
- Comparisons between phenomena in terms of position (spatial) and over time (change).
- Accurate and careful descriptions.

Again, these are indicators of the *quality* of children's observations. To assess these teachers need to develop specific scoring rubrics based on the nature of the activity and the emphasis on the children's observations in the teaching-learning process.

Assessing for Science Attitudes

The purpose for assessing attitudes in children's journals is not to reward the expression of positive attitudes or to penalize the expression of negative ones, but to reward children for representing their feelings and attitudes about the science experience using written language or drawing. In order to assess children's attitudes, teachers can analyze the journal for expressions of excitement, interest, and curiosity. In addition, teachers may look for reflections on cooperation, tolerance of ambiguity and persistence, and resourcefulness.

Table 10. Sample Checklist for Assessing the Science Processes in Children's Journals

Science processes	Excellent complete and accurate	Acceptable partially complete and accurate	Unacceptable incomplete and inaccurate	Not observed
Clearly states question(s)				
Clearly states prediction(s)				
Organizes and neatly presents data				
Represents data in chart or graph				
Describes procedures				
Explains the use of materials and equipment				
Identifies all variables				
Controls variables				
Bases conclusions on evidence				

Assessing Children's Drawings

Children's journal drawings may be analyzed to assess all of the science domains. Specific performance indicators for children's drawings reflect:

- Appropriate and accurate sequences.
- Detail, scale, and accuracy.
- Actual objects versus stereotypic drawings.
- Relationships between objects and realistic positions.
- Care and neatness.

Constructing Scoring Rubrics for Assessing Children's Journals

Scoring rubrics for journals need to assess the science domains based on performance indicators. The particular scoring rubric developed will depend on the nature of the science activity and on the instructional goals and values identified for the activity. Although there is no single generalizable scoring rubric, there are guidelines for developing reliable and valid scoring rubrics. Teachers first determine whether they wish to assess the child's journal as a whole product, giving a single score, or as a set of individual components, giving multiple scores. In the first case, a holistic scoring rubric is needed and in the second case an analytic scoring rubric is required.

Regardless of the scoring system selected, the rubric must be built upon the science learning valued by the teacher and on the instructional goals for the science experience. After clearly stating these instructional goals and values, determine which indicators will accurately assess children's journals in this. After identifying the indicators needed, define the criteria that will accurately assess the child's performance for each indicator. After identifying the criteria, describe the possible performance levels based on the criteria, using either an analytic or a holistic format. The performance levels should be clearly defined so that they differentiate children's abilities. The performance levels then serve as the scoring rubric used to assess children's journals. Next, the scoring rubric is revised based upon children's work. The planning matrix presented in Table 11 may assist in the development of the scoring rubric for assessing children's science journals. In summary the steps are as follows:

1. Identify the instructional goals and values.
2. Determine the indicator(s) for each instructional goal or value.
3. Define the criteria for each indicator.
4. Describe the performance levels (scoring rubric) to differentiate children's abilities.
5. Revise the scoring rubric based upon children's work.

Suggestions for Successfully Assessing Children's Journals

If developing a holistic scoring rubric, three to four simple and easy to use performance levels should be developed. Scoring rubrics should be revised based upon children's journal entries to ensure that the rubric aligns with children's abilities and differentiates performance levels. Limit each journal assessment to three or four goals and two or three indicators per goal. This not only ensures that multiple science domains are assessed through multiple indicators, but that the assessment is efficient and practical to use. Rotate the goals and indicators so that each is assessed in multiple activities over time. This develops a picture of the child's abilities throughout the school year, and assures manageability.

Scoring rubrics should be shared with parents, with other teachers, and with school administrators to keep them informed about this use of children's journals and the assessment procedure. Keep examples of children's work as evidence of performance. Share scoring rubrics with the children before they engage in the science activity and journal writing; this helps children understand the expectations, and will guide them in completing the task.

Table 11. Planning Matrix for Developing Scoring Rubric to Assess Children's Science Journals

Step 1 **Instructional Goal**	Step 2 **Indicators**	Step 3 **Criteria for Indicators**
1.	1.	1.
		2.
	2.	1.
		2.
	3.	1.
		2.
2.	1.	1.
		2.
	2.	1.
		2.
	3.	1.
		2.
3.	1.	1.
		2.
	2.	1.
		2.
	3.	1.
		2.
Step 4 **Performance Levels** **Based on Criteria**	3. 2. 1	

Closing Thoughts on Children's Journals in Science Teaching and Assessing

The current curricular and instructional push to use children's journals in the teaching-learning-assessing process has sometimes resulted in their superficial use in science classrooms. Often, curricular materials and instructional documents provide the classroom teacher with insufficient background for using children's journals. This can result in confusion and a sense of obligation rather than in a belief in the value of children's journals (Parsons, 1994). Under these circumstances, the use of journals has little consequence, and leaves an acrid taste in the mouths of children. Simply using science journals for the sake of following a curricular plan is counterproductive. Using children's science journals in the teaching-learning process without assessing them is pedagogical malpractice, ignoring the importance of children's cognitive and verbal efforts to make sense of science phenomena.

ACKNOWLEDGEMENTS

Portions of the children's journal section were published in Shepardson, D.P. & Britsch, S.J. (1997). Children's science journals: Tools for teaching, learning, and assessing. *Science and Children*, 34(5), 13-17 & 46-47.

REFERENCES

American Association for the Advancement of Science (1993). *Benchmarks for Science Literacy.* New York: Oxford University Press.

American Association for the Advancement of Science (1998). *Blueprints for reform: Science, mathematics, and technology education.* New York: Oxford University Press.

Bell, B., Osborne, R., & Tasker, R. (1985). Finding out what children think. In R. Osborne & P. Freyberg (Eds.), *Learning in science: Implications of children's science* (pp.151-165). Auckland, New Zealand: Heinemann Publishers.

Britsch, S. J. & Shepardson, D.P. (1996). *Science journals: Tools for the construction of understanding.* Paper presented at the annual meeting of the American Educational Research Association, New York, NY.

Broadfoot, P. & Towler, L. (1992) Self-assessment in the primary school. *Educational Review*, 44(2), 137-51

Dana, T.M., Lorsbach, A.W., Hook, K., & Briscoe, C. (1991). Students showing what they know: A look at alternative assessments. In G. Kulm & S.M. Malcom (Eds.), *Science assessment in the service of reform* (pp. 331-337). Washington, DC: American Association for the Advancement of Science.

Doris, E. (1991). *Doing what scientists do: Children learning to investigate their world.* Portsmouth, NH: Heinemann.

Elstgeest, J., Harlen, W., & Symington, D. (1985). Children communicate. In W. Harlen (Ed.), *Primary science: Taking the plunge* (pp. 92-111). Oxford, England: Heinemann.

Freedman, R.L.H. (1994). Open-ended questioning: A handbook for educators. Menlo Park, CA: Addison-Wesley.

Harlen, W. (1983). *Guides to assessment in education.* Hong Kong, Macmillan.

Harlen, W. (1985). Helping children to plan investigations. In W. Harlen (Ed.), *Primary science: Taking the plunge* (pp. 58-74). Oxford, England: Heinemann.

Harlen, W. (1988). *The teaching of science.* London, England: David Fulton Publishers.

Harlen, W. (1993). *Teaching and learning primary science.* London, England: Paul Chapman Publishing.

Hodson, D. & Brewster, J. (1985). Towards science profiles. *The School Science Review*, 67, 231-40.

Ollerenshaw, C. & Ritchie, R. (1997). Primary science making it work (2nd edition). London, England: David Fulton Publishers.

Parsons, L. (1994). *Expanding response journals in all subject areas.* Portsmouth, NH: Heinemann.

Shepardson, D.P. (1997). Of butterflies and beetles: First graders' ways of seeing and talking about insect life cycles. *Journal of Research in Science Teaching*, 34(9), 873-889.

BRENDA MAIN

9. ASSESSING CHILDREN'S SCIENCE LEARNING AND PROCESS SKILLS IN THE ELEMENTARY CLASSROOM

Today's classroom teacher must meet the needs of a very diverse group of students. *Science for all Americans* (American Association for the Advancement of Science [AAAS], 1990) emphasizes inclusiveness, assuming that no individual or group is excluded from an opportunity to become science literate, and that no student is presumed unable to become science literate. Students have a better opportunity to develop into well-rounded and scientifically literate individuals if teachers use a variety of instructional methods and a variety of assessments. The aim is to provide each student an opportunity to feel comfortable enough to strengthen and develop understandings and skills.

In this chapter I will share my experiences and ideas about science assessment in elementary classrooms. I will specifically address inquiry-based teaching and assessing, drawing examples from a sink and float assessment task. This example illustrates how assessment and instruction can be closely aligned to diagnose children's understandings, inform pedagogy, and evaluate student performance. I will also discuss the use of practical task assessment stations, student interviews as assessment, observational checklists, and informal assessment.

I returned to teaching third grade after eleven years of teaching science to fourth- and fifth-grade students. Third grade students need experience in planning and conducting simple investigations, using the science process skills, and engaging in scientific inquiry. The National Science Education Standards (National Research Council [NRC], 1996) indicate that scientific inquiry for third-graders should involve students in:

- Raising questions and answering their questions through observations.
- Planning and conducting simple experiments to answer questions.
- Using materials and simple equipment to collect data.
- Communicating both their investigations and explanations to others.

In addition, one of my personal goals is to encourage my third-grade students to love science as I do. The classroom environment, teacher enthusiasm toward science, and the science experiences and social interactions that comprise the classroom culture (Tishman, Perkins, & Jay, 1995) all influence students' dispositions toward science. A positive disposition is best developed through

149

Daniel P. Shepardson (ed.), Assessment in Science, 149—161.
© 2001 *Kluwer Academic Publishers. Printed in the Netherlands.*

inquiry-centered science teaching, in my opinion. This has numerous benefits: it actively engages children in learning science; it relates science to the child's world; it promotes collaboration; it accommodates different learning styles, and it links concepts and skills to actual materials and phenomena (National Science Resources Center [NSRC], 1997).   I want my students to enjoy science while they develop their science process skills, the processes of observing, comparing, communicating, investigating, predicting, recording and using data, and inferring.   Children do not develop these skills in a vacuum, however, but in the context of the science investigation or activity. Assessment of children's performance of these skills needs to occur in the context of investigations and activities.   In this way, assessment reflects the children's classroom experiences.

Although multiple-choice test have a place in the elementary classroom, they cannot assess children's inquiry abilities or their capability to use the science process skills. I have developed and implemented assessments that are stand-alone practical tasks and that are embedded into instruction. Embedded assessments are activities that students engage in; they serve as both formative and summative assessments.   Embedded assessments enable teachers to obtain and record information about students' (NSRC, 1997) science process skills as students engage in the instructional activity.

SINK AND FLOAT: AN ASSESSMENT TASK EXAMPLE

My school corporation's curriculum guide states that a nine-week period of time be devoted to each of the following: earth science, physical science, life science, and health.   Sinking and floating is a portion of the third grade physical science curriculum.   It is imperative that science experiences for children integrate the science process skills and build on the curiosity of children as well as the content of the curriculum (Frederick & Cheesebrough, 1993).   I wanted to create an inquiry experience that would motivate third graders and engage them in using and developing their science process skills.   I utilized the *Benchmarks for Science Literacy* (AAAS, 1993) to assist me in identifying the appropriate inquiry standards for this age student.

In order to help my students develop a better understanding of sinking and floating I felt they would benefit from an inquiry experience with objects. Based on the science education research literature, however, I realized that my students would have conceptual difficulty understanding density and buoyant force. Therefore, I expected my students to understand that objects sink and float based on their size, shape, and weight, that all heavy objects do not sink and that all small objects do not float.   A major focus of my science curriculum is also to help students make real world connections to the investigations they experience in class.   In this case, students by understanding that objects can be changed to make them sink or float or by relating sinking and floating to boats and rafts.

The research literature informed me that young children tend not to differentiate weight and density; instead, they include weight and density in the general notion of heaviness (Smith, Carey, & Wiser, 1984). Piaget (1973) found that weight and density develop as children attend to other viewpoints and that by the age of nine children begin to relate the density of one object to that of another. For example, a rock sinks because it is heavier than the water; a feather floats because it is lighter than the water. Most children believe objects float because they are light, that the length of an object determines whether it will sink or float, that long objects would sink, and the water displaced by a floating object would not be related to the density of that object (Biddulph & Osborne, 1984). Children's ideas about buoyancy include the notions that: holes in an object affect its ability to float; objects become heavier and sink if water flows inside them, water pressure pushes either upwards or downwards, and objects with air inside will float (Grimellini, Gandolfi, & Balandi, 1990).

Existing curricular guides and the research literature helped me create an inquiry activity on sinking and floating. As part of this activity, students would weigh each object and observe the objects shape and size to look for patterns in those objects that sink and float. Requiring students to weigh the objects aligned the activity with the *Benchmarks for Science Literacy* (AAAS, 1993) where students are to ". . . use numerical data in describing and comparing objects and events" (Habits of Mind, Communication Skills, AAAS, 1993, p. 297). A critical element is the careful selection of objects for students to manipulate. Objects must be the same in size and shape, but different in weight; some objects that are large and heavy that float and objects that are small that sink.

I created both analytic (Table 1) and holistic (Table 2) scoring rubrics to assess students in this activity I was still learning about scoring rubrics and wanted to find out which one would be the most comfortable for me to use and which would be more efficient and accurate in assessing student performance. I found that constructing the scoring rubric prior to initiating the activity better focused the learning aims for that activity. I also share the scoring rubrics with my students before the activity. I have found that this guides my students because the scoring rubric makes explicit the criteria and expectations for completing the activity as an assessment task.

As an introduction to the sink and float activity, I conducted a whole class discussion while demonstrating that items either sank or floated. Students first made predictions and explained their predictions for each object prior to its demonstration. Students recorded their predictions and explanations in their science journals. The assessment task required students to keep a journal that described their observations and distinguished actual observations from ideas about what was observed (Benchmarks, Habits of Mind, Manipulation and Observation, AAAS, 1993, p. 293). Students were also required to ". . . offer reasons for their findings and consider reasons suggested by others" (Benchmarks, Habits of Mind, Values and Attitudes, AAAS, 1993, p. 286). This predict-explain-observe strategy is a variation of the prediction-observation-explanation technique described by White and Gunstone (1992). Such prediction techniques reveal students' conceptual understandings, promote discussion of ideas, motivate students, and demonstrate to students that they already have ideas about the science topic (White and Gunstone, 1992). The prediction activity served as both an instructional activity to engage students in thinking about sinking and floating, and as a diagnostic assessment of the children's ideas and understandings.

Table 1. Analytic Scoring Rubric for the Sink-and-Float Activity

Performance Category	Description of Performance	Score
Observing	Student notes size, shape, and weight of each object and whether it sinks or floats.	3 2 1 0
Comparing	Student compares objects that sink and float based on their size, shape, and weight.	3 2 1 0
Communicating	Student clearly presents predictions, observations, results, and inferences.	3 2 1 0
Investigating	Student records steps in the investigation, follows procedure, and draws conclusions.	3 2 1 0
Predicting	Student makes prediction for each object with explanation.	3 2 1 0
Recording data	Student records object and information on size, shape, weight, and whether it sinks or floats.	3 2 1 0
Inferring	Student draws conclusions about how size, shape, and weight of an object influence sinking or floating.	3 2 1 0

Scoring scheme: 3 pts.   Complete and accurate
                  2 pts.   Complete but some inaccuracies or errors
                  1 pt.   Not complete and with inaccuracies or errors
                  0 pts.   Not attempted

Table 2. Holistic Scoring Rubric for the Sink-and-Float Activity

Performance Level	Description
3: Exceptional Performance	Student observations, comparisons, communication, investigation, predictions, data recordings, and inferences are complete, clear, and accurate.
2: Average Performance	Student observations, comparisons, communication, investigation, predictions, data recordings, and inferences are complete and clear, but contain some inaccuracies or errors.
1: Needs Improvement	Student observations, comparisons, communication, investigation, predictions, data recordings, and inferences are incomplete and/or not clear, containing many inaccuracies or errors.
0: Not Attempted	Student does not complete task

As a diagnostic assessment, it revealed children's ideas and understandings based on their predictions and explanations about why an object would sink or float. This was an informal assessment because children's understandings were not graded, but simply recorded for the purpose of guiding and structuring the inquiry investigation that followed. Most day-to-day assessments in elementary classrooms are informal, a seamless part of the teaching-learning process. For the most part, this assessment is intuitive; it is initiated at the spur of the moment, and it is dependent on the situation at hand (Wragg, 1997). Informal assessments, however, may be pre-planned and focused. I find that although I consistently conduct unplanned or spontaneous informal assessments, I gain the most insight into children's understandings and pedagogy by planning my informal assessments (see section below for more about informal assessment). Therefore, as an informal assessment, this prediction activity provided me the opportunity to learn the extent of my students' prior knowledge about sinking and floating; this enabled me to adjust the inquiry experience based on these understandings. It also provided me with a base for comparing the students' developing understandings about sinking and floating.

For me, the prediction activity also models for students the scientific process of making a good prediction; that what is important is that you base your prediction on evidence. Students can make better predictions if they use the results from a previous test or experience to make new predictions. This is as important as obtaining the "right answer." Further, this process models the difference between a prediction (based on prior knowledge, data, or information) and a guess. I consider this to be a prerequisite to creating a hypothesis in science experiments.

For the sink-and-float activity I developed a scoring system that assessed children's use of the basic science process skills (i.e., predicting, observing, record keeping, and communicating). This pedagogical-assessment activity also required students to transfer their recorded data into a graphable product. This helped to set the stage for the later development of data tables or graphs. In third grade, students began using science journals to record each step of the scientific process. It was essential for them to learn the importance of keeping accurate, detailed observations. If these skills are developed in elementary school it will be easier for students to write detailed laboratory reports in middle and high school. This component of the assessment task is consistent with the *Benchmarks for Science Literacy* (AAAS, 1993):

> Results of scientific investigations are seldom exactly the same, but if the differences are large, it is important to try to figure out why. One reason for following directions carefully and for keeping records of one's work is to provide information on what might have caused the differences (The Nature of Science, Scientific Inquiry, AAAS, 1993, p. 11).

> Keep records of their investigations and observations and not change the records later (Habits of Mind, Values and Attitudes, AAAS, 1993, p. 286).

The second portion of the assessment task provided an opportunity for students to transform information gained in the first part of the inquiry investigation into a simple investigation of their own. Students were required to record their procedures and to make drawings in support of their explanations. This second portion of the assessment task is in accordance with the *Benchmarks for Science Literacy* (AAAS, 1993) where students are to ". . . write instructions that others can follow in carrying out a procedure. Make sketches to aid in explaining procedures or ideas" (Habits of Mind, Communication Skills, AAAS, 1993, p. 297). Using what one has learned to solve related problems helps students connect classroom experiences to their lives and to connect science to all aspects of life. Asking questions and making connections are essential to a successful science program.

This group of third graders had little trouble in predicting, recording, and then testing and recording each object provided. Those children who were obtaining unexpected results stated this in the journal portion of the investigation. Although I felt the instructions for the inquiry investigation were clear it became apparent that I had not been as specific as I had thought. The students had no difficulty in identifying five objects that sank and five that floated and they transferred this information to their response sheets. The problem arose when children started to draw the objects on the response form provided. Students began to ask whether I could tell what their picture represented. I realized that I had only asked the students to draw the objects in the container and not to label what they were. A verbal instruction soon rectified this error. Most of the class dud this, but three students depicted objects up in the air above the container of water. I found this very confusing and at first did not understand why they drew the objects in the air. I later learned that the visualization of three-dimensional objects on a flat surface is developmental. All of my eight- and nine-year old students may not have been developmentally able to handle a three-dimensional container. From a Piagetian perspective, some of my students had not yet developed projective structures that would enable them to coordinate and reference their drawing with their actual observations (Piaget & Inhelder, 1969). Next time I will provide only a rectangle to represent the container and let the students draw the water as well as the objects.

In the second part of the activity I wanted the students to not only investigate, but also to record their results in complete sentences. The students did not try as many ways to make the Styrofoam sink as I had hoped. Once they had a successful method, their investigation came to a stop. I had hoped that by being more specific in what I asked the students to do I would rectify this situation. As a result of these discoveries, I redesigned the student response sheet. I hope to have better results with this inquiry investigation with these revisions. A good inquiry investigation or assessment tool does not always happen on the first attempt and none should ever be considered perfect; scientific information changes and the ability level of the students can differ greatly from year to year. The availability of materials is also a factor that may cause a need for change. Change should not be looked upon as, I am not any good at this, but as I am willing to improve upon this for the benefit of my students.

As a result of these discoveries I once again redesigned the student response sheet. I re-instated the "notes" portion to the record-keeping sheet. I replaced some of the objects to be tested. Based on a few more experiences with scoring rubrics, I decided to add an "expectation" section to the activity. I hope these revisions will have better results, but if I need to revise the assessment task further I will not hesitate to work on it again. After scoring this inquiry investigation with both the holistic and analytical scoring rubrics, I decided I preferred the analytic scoring rubric. I found it easier to score student responses this way, and I felt I was able to recognize those students who did well on some components of the activity, but who needed improvement in other areas. The holistic scoring rubric resulted in lower scores and was more difficult to apply to students' work. I redesigned the analytic scoring rubric to meet the needs of the redesigned assessment task and student response sheet.

OTHER ASSESSMENT METHODS

In this section, I wish to discuss other assessment methods appropriate for use in elementary schools: Informal assessments, student interviews as assessment, practical task assessment stations, and observational checklists. The use of these methods in conjunction with traditional assessment formats (e.g. selected response and constructed response) and the previously mentioned pedagogical-assessment task (sink-and-float) bases assessment practice on multiple assessment formats that assess multiple student traits and science domains. These assessment methods can be used throughout the unit of study and over the academic year, providing multiple measures of student learning.

Informal Assessments

The most common method of informal assessment is oral questioning. Because two-thirds of teacher questions are associated with assessment, it is imperative that careful thought be given to the way in which questions are asked and the use of student responses (Wragg, 1997). Questions such as: "Why do you think . . .? What can you tell me about . . .? How do you know . . .?" provide students with opportunities to share their thinking and understandings (Bell, Osborne, & Tasker, 1985). For example, I might have asked students, "Why do you think the Styrofoam ball will float? How do you know the wooden block will sink? What can you tell me about the wooden block?" By asking these questions, I gain insight into my students' ideas and understandings. I would then use their responses to adjust my teaching by providing different objects or by asking questions that guide children's observations and activity; Elstgeest (1985) calls these productive questions.

Informal assessments can also take the shape of practical tasks. Instead of demonstrating the sink-and-float objects and having students record their predictions and explanations, I could have provided the objects and asked students to group them into "sink" and "float" categories. Students would be required to explain why all the objects in the "sink" group would sink and why all the objects in the "float" group would float. I could then ask questions specific to these groupings, with students' explanations as the important indicator of their understandings. I could again use productive questions to focus students' observations and guide the actual testing of predictions. In both of these examples, informal assessment diagnoses children's understandings and informs pedagogy; it does not evaluate or grade students.

Interviews as Assessment

During inquiry activities, I often use oral interviews as a formative assessment method. This assesses students' understanding of the inquiry activity, their existing conceptual understandings, and the understandings they are constructing from the inquiry experience. I have used oral interviews as both formal and informal assessments, but in both cases I have developed or planned the interview questions prior to the inquiry activity. When I use oral interviews as a formal assessment I construct a matrix recording the students names and a list of two or three questions (Table 3). By completing the interview matrix, I have a record of student performance that I can use as assessment evidence during parent-teacher meetings, in determining student grades, or in monitoring student progress. I also record anecdotal notes about student responses. I record my assessment of student response as:

+ (Completely accurate)
√ (Somewhat accurate)
− (Mostly inaccurate)

Table 3. Example Interview Assessment Record Matrix

Interview Questions	Students				
How would you explain why an object sinks (floats)?					
What have you observed about the objects that supports your idea about why objects sink (float)?					
What have you learned about sinking and floating?					

+ = Completely accurate; √ = Somewhat accurate; − =Mostly inaccurate

Drawing from Bell (1995), I have used the following question categories to guide my thinking about and development of oral interview questions:

- What are students asking?
- What are students' explanations?
- Are students relating what they already know to new ideas?
- Are students constructing meanings to that intended?
- Can students provide a reason/evidence to support their ideas?

I have aligned several sample questions I use based on Bell's categories:

- What are students asking? I sometimes ask students "If you were going to ask a question about this experience what would you ask and why?"
- What are students' explanations? I would ask students questions like: "How would you explain this investigation to another teacher or student? How would you explain the result of this investigation?"
- Are students relating what they already know to new ideas? Here I would ask: "How is this the same or different from what you already know? How does this relate to what you already know? How does this compare to your prediction?"
- Are students constructing meanings that relates to the task? To determine this I would compare my expectations to student responses.
- Can students provide a reason and/or evidence to support their ideas? Questions I might ask within this category are: "What is your evidence for saying that? What data or results support what you are saying?"

Assessment Stations

At times I assess students through the use of several practical tasks, in a station format, where students rotate through each station completing the practical task. Each station emphasizes a particular science domain or concept so that I can assess several domains or student traits related to the science topic. I often observe students using an observational checklist to assess their performance, and rely on students' products produced at the other assessment stations. For example, in the assessment of students' performances in a unit on Earth materials, I might have three practical task assessment stations. The first task might require students to perform several geological tests to determine which rocks are sedimentary. In this task, students would use the provided materials to conduct various tests of the rock samples. Students would be provided a response form where they record which rocks were sedimentary and the evidence supporting their thinking. I would also observe students as they conducted the investigation to determine whether they conducted the same tests on all rock samples in a similar manner. The second station might require students to determine the hardness of the rock samples and order the rock samples based on hardness. The student response form would require students to list the rock samples in this order, providing evidence from the hardness test to support their rankings. The third assessment station might ask students to identify three rock samples using the resources that describe and classify the different rock types. Students record the rock types and the evidence to support their identifications.

In conjunction with the assessment stations I also incorporate open response tasks and traditional assessment approaches such as multiple-choice and fill-in-the-blank questions that children complete when they are not working at the assessment stations. This summative assessment approach incorporates multiple measures in diverse formats, assessing multiple student traits and science domains. The assessment stations align with my instruction and require students to make observations, use information, record data, communicate their findings and ideas, conduct simple investigations, and make inferences. Based on my experiences I share these additional suggestions for successfully using practical task assessment stations:

- Avoid the use of materials or equipment that children are not familiar with; this adds another dimension to the practical task—an uncontrolled variable. The assessment task should use materials and equipment children have had experience with.
- Be sure each station is clearly identified and separated from the other so that children do not mix the materials and equipment from each practical task.
- All necessary materials and equipment should be located at each station. Children should not have to go to another location to complete the task.
- Provide sufficient time for children to complete the practical task. Children need time to manipulate materials and to use the equipment; otherwise, the task becomes an assessment of children's speed instead of an assessment of their science learning. Remember haste makes waste and mistakes. I allow about 15 minutes per station.
- Limit the number of assessment stations to three. Set up at least two stations of each task. This allows six children to work on the practical tasks at a time. Be sure children rotate through the stations in order to avoid confusion. I set up the stations around the outside of the room.

Baxter and Shavelson (1994) found that practical task assessments, when compared to other assessment methods, provide a better picture of what students know and can do in science. This result was confirmed by the research of Sugrue, Webb, and Schlackman (1998) who also found that if the written portion of the assessment is conceptually similar to the practical tasks at the assessment stations, students who complete the practical tasks first might score higher on the written portion of the assessment.

Observational Checklists

I have successfully used observational checklists to assess students' performances during inquiry-based activities. Observational checklists provide information on students' abilities to perform the science process and inquiry skills within the context of instruction. The four key ingredients I have found in designing functional classroom observational checklists are the following:

- Limit the number of science process skills or behaviors (indicators) observed at any one time to three or four.

- Rotate the science process skills or behaviors observed so that over time you observe different science process skills and behaviors multiple times throughout the academic year.
- For each inquiry investigation identify specific science process skills or behaviors to observe.
- Do not be afraid to ask students questions in order to clarify an action. By asking a question about their action you can better understand their behavior. Why they did what they did.

In the sink-and-float investigation, I might wish to assess students' ability to record data. I must first decide whether I will assess students' ability to record data throughout the academic year. If I decide that I am not going to assess students' ability to record data throughout the academic year, then it may be meaningless to assess students' ability to record data in the current activity. It would be more informative to assess a science process skill that I plan to assess throughout the academic year; this would provide more and better information on students' progression in the ability to perform the science process skill. To construct the observational checklist, I identify the specific actions involved in recording data, keeping in mind my students' developmental level and the experiences they have had in recording data. These identified actions become the indicators of student performance (what I will observe). For the sink and float investigation I would expect students to:

- Record the name of each object
- Record the weight of each object
- Record the size of each object
- Record the shape of each object
- Record whether the object sinks or floats
- Accurately and methodically record the data for each object

I rate the student's performance on each observational category based on four levels: exemplary performance, average performance, needs improvement, and not observed. Observational checklists embed assessment in instruction, and evaluates and diagnoses students' performance capabilities. Each observational checklist may be placed in student's portfolio as evidence of progress in the ability to perform the science process and inquiry skills. Analyzing the student's portfolio then provides classroom-based evidence of their ability to perform the science process and inquiry skills (NRC, 1996). By assessing science process and inquiry abilities during instructional activities we send a message to students that, as teachers, we value the ability to engage in science.

CONCLUDING THOUGHTS ABOUT ASSESSMENT IN SCIENCE

Alternative assessment is the multidimensional measurement of student performance, utilizing multi-trait, multi-source techniques as indicators and evidence of student performance. By using a variety of assessments, teachers help students to develop different ways of accomplishing various tasks. This will help each to become a better problem-solver in life and to understand that all individuals have various learning styles. In this way, students can develop their strengths and improve upon their weaknesses. Assessments also enable teachers to diagnose student needs and to see what degree of growth children have obtained. There are many ways to assess students and their products. In creating assessments teachers must decide what it is they want their students to accomplish within the assessment task. These assessment tasks may take the form of problem-solving situations, inquiry investigations, and tasks that combine reading and writing. Scoring rubrics may be used to assess student products and the science processes. Scoring rubrics also help teachers to be consistent and to establish credibility in their assessment process.

In teaching, it is essential that we remain open-minded. We must be life long learners and remain current in all areas of our field. Change is needed to meet the needs of today's students in all areas of education, not just science. One way to help students develop and grow is by using a variety of teaching and assessment methods. Each of us should challenge ourselves to become assessment literate; however, no one should attempt too many changes in assessment at once. Each of us should try out and then fine-tune our assessment methods with the understanding that real change takes time.

REFERENCES

American Association for the Advancement of Science (1990). *Science for all Americans.* New York: Oxford University Press.

American Association for the Advancement of Science (1993). *Benchmarks for Science Literacy.* New York: Oxford University Press.

Baxter, G.P. & Shavelson, R.J. (1994). Performance assessments: Benchmarks and surrogates. *International Journal of Educational Research, 21,* 279-298.

Bell, B. (1995). Interviewing: A technique for assessing science knowledge. In S. Glynn & R. Duit (Eds.), *Learning science in the schools: Research reforming practice* (pp. 347-364). Mahwah, NJ: Lawrence Erlbaum Associates.

Bell, B. Osborne, R., & Tasker, R. (1985). Finding out what children think. In R. Osborne & P. Freyberg (Eds.), *Learning in science: Implications of children's science* (pp. 151-165). Auckland, New Zealand: Heinemann Publishers.

Biddulph, F. & Osborne, R. (1984). Pupils' ideas about floating and sinking. Paper presented at the Australian Science Education Research Association Conference, Melbourne, Australia.

Elstgeest, J. (1985). The right question at the right time. In Harlen, W. (Ed.), *Primary science: Taking the plunge* (pp. 36-46). Portsmouth, NH: Heinemann.

Frederick, A. & Cheesebrough, D. (1993). Teaching mathematics. New York: McGraw-Hill.

Grimellini, N., Gandolfi, E., & Pecori, B. (1990). Teaching strategies and conceptual change: Sinking and floating at elementary school level. Paper presented at the Annual Meeting of the American Educational Research Association, Boston, MA.

National Research Council (1996). *National science education standards.* Washington, DC: National Academy Press.

National Science Resources Center (1997). *Science for all children: A guide to improving elementary science education in your school district.* Washington, DC: National Academy Press.

Piaget, J. (1973). *The child's conception of the world.* London, England: Paladin.

Piaget, J. & Inhelder, B. (1969). *The psychology of the child.* New York: Basic Books, Inc.

Smith, C., Carey, S., & Wiser, M. (1984). A case study of the development of size, weight, and density. *Cognition*, 21(3), 177-237.

Sugrue, B., Webb, N., & Schlackman, J., (1998). *The interchangeability of assessment methods in science*. CSE Technical Report 474. Los Angeles, CA: Center for the Study of Evaluation, National Center for Research on Evaluation, Standards, and Student Testing, Graduate School of Education & Information Studies, University of California.

Tishman, S., Perkins, D., & Jay, E. (1995). *The thinking classroom: Learning and teaching in a culture of thinking*. Boston, MA: Allyn and Bacon.

White R. & Gunstone, R. (1992). *Probing understanding*. London, England: The Falmer Press.

Wragg, T. (1997). *Assessment and learning: primary and secondary*. London, England: Routledge.

MARILYNN EDWARDS

10. PEDAGOGICAL-ASSESSMENT ACTIVITIES IN A MIDDLE SCHOOL LIFE SCIENCE CLASSROOM

Alternative assessment improves the science education of students by aligning curriculum, instruction and assessment. The results of alternative assessments may be used to plan the curriculum and guide instruction, as well as to evaluate students. I use inquiry activities as my main vehicle for teaching because of the interest seventh graders have in "doing science." Inquiry-based activities allow students to observe, interpret, and construct knowledge for themselves in a manner that cannot be accomplished with traditional textbook based teaching methods. Students must get their hands on science to get their minds on science. In this way, I use inquiry-based activities as assessment tasks that align my assessment practice with my instructional approach.

In this chapter I will share an example of a pedagogical-assessment activity, including student work examples or responses to the assessment task, the scoring guide I use to assess student performance, and my thinking about assessment in the seventh grade. I will also share a brief example of a pedagogical-assessment activity that serves as a summative assessment task and discuss the notion of standards-based assessment. The ultimate goal of this chapter is to demonstrate how instructional activities may be used as assessment tasks: pedagogical-assessment activities. If classroom assessment is to become a more powerful tool in measuring student growth, it must move from the "end-of-the-week, end-of-the-chapter, end-of-the-unit" procedure to ongoing formative assessments (Burke, 1992, p.49).

I evaluate student performance with assessment tasks that allow for open-ended responses, because I want my students to reflect on what they are observing and to communicate their findings in a variety of ways. One of the alternative assessment tasks I use, "What is the Relationship between Abiotic and Biotic Factors?" demonstrates the interaction of living and nonliving factors in a dynamic manner. This task is also an instructional activity because it enables students to learn about the interactions between abiotic and biotic factors. This approach embeds assessment in instruction creating pedagogical-assessment activities. These may be characterized according to attributes that Baron (1991) identified for enriched performance assessment tasks. These include:

- Involves students in sustained work, often taking several days to complete.

Daniel P. Shepardson (ed.), Assessment in Science, 163—179.
© 2001 *Kluwer Academic Publishers. Printed in the Netherlands.*

- Emphasizes the essential concepts of the discipline, requiring students to apply concepts to explain situations and events.
- Integrates both science content and processes in completing the task.
- Requires students to explain and define their predictions.
- Supports interpretations and conclusions with evidence.
- Creates products that demonstrate understandings.

In addition, pedagogical-assessment activities consist of an explicitly stated scoring system that guides students in the completion of the task. Pedagogical-assessment activities serve to both teach and assess students, providing students with an opportunity to learn from the assessment activity (National Research Council [NRC], 1996).

OVERVIEW OF THE ASSESSMENT TASK

This pedagogical-assessment activity (i.e., "What is the Relationship between Abiotic and Biotic Factors?") is second in the thematic set "Energy in a Living System" that I teach. In this assessment task, students observe four jars containing different combinations of bromothymol blue, plants, and snails for four days. The students are to record their observations of the interactions between the living and nonliving factors of the environment, and describe their observations in both drawings and words. I ask them to both draw and explain their observations because this combination provides an outlet for the visual as well as the verbal learner. Interdependency is a difficult concept to explain with one-word responses; therefore, if I am to learn what students know, I must provide them the opportunity to fully express themselves.

Students are to reflect on their observations before they provide reasons for the changes they observe; I want them to interpret the situation and then represent their mindscapes on paper. They are to *explain* what they think is happening in this microenvironment, not to guess. An important aspect of scientific literacy is the capacity to engage in scientific ways of thinking (American Association for the Advancement of Science [AAAS], 1990). I expect my students to become responsible for their own learning as they conduct the activity by supporting their ideas and questions and by pursuing their own thinking. Constructivist theory suggests that students must be responsible for their own learning (Brooks & Brooks, 1993); this responsibility for learning is a trait we value in scientifically literate individuals.

The assessment task was developed in alignment with national standards for students' science learning. The *Benchmarks for Science Literacy* (AAAS, 1993), "Habits of the Mind," state that a student should be able to make an observation that is distinguishable from speculation. At the end of the fifth grade, most students should be able to communicate their observations, the relationships they observe, and their interpretations. The assessment task requires students to practice these skills, routinely performed in my science classroom. "The Living Environment" section of the *Benchmarks for Science Literacy* (AAAS, 1993) indicates that students need guided practice to understand the interdependency of organisms, and that by the end of the eighth grade they should know that:

> Two types of organisms may interact with one another in several ways ... Relationships may be competitive or mutually beneficial. Some species have become so adapted to each other that neither could survive without the other (The Living Environment, Interdependence of Life, AAAS 1993, p. 117).

This means that seventh grade students should understand that the snail and plant are interdependent on each other within this microenvironment. The plant produces oxygen that the snail needs to breath and at the same time the snail respires providing carbon dioxide that the plant uses during photosynthesis.

When I use this assessment task my instruction matches the curriculum, and the assessment provides feedback for me to effectively judge whether students understand the interdependency between biotic factors of the environment. Students use words and drawings to both demonstrate and communicate their understanding and knowledge of ecological concepts that they will use in the summative assessment for the unit: increasing the life span of the organisms in their bottle ecosystem.

DEVELOPING THE ASSESSMENT TASK

I developed this pedagogical-assessment activity on biotic and abiotic factors of the environment because I found the textbook lacking in activities. From experience, I knew that not all of my students really understood the interaction between living and nonliving components of the environment. They could repeat the definition of the terms, but many did not demonstrate an understanding of this concept when they chose organisms for their bottle ecosystem. In particular, students did not include plants as a necessary part of the bottle ecosystem. From another pedagogical-assessment activity, I discovered a gap in students' understandings that for years had remained invisible because of my teaching and assessment practices. I have come to realize that alternative assessments provide a deeper insight into students' understandings; this in turn, better informs my pedagogy and this results in changes in my teaching. Traditional assessments, such as fill-in-the-blank and selected response questions (e.g. multiple-choice, true and false, and matching) did not show me whether students really understood the concepts behind the activities. Students could recite the definition of the concept, but could not confirm that understanding when conducting an inquiry investigation requiring them to apply the understanding.

In an attempt to find the reasons for this conceptual gap, I developed and implemented another alternative assessment task as a diagnostic tool to identify students' prior knowledge about biotic and abiotic factors. I constructed an assessment task in an open-ended response format that contained five prompts based on a sketch of a woodland area. The prompts for the assessment task were:

- List the living things found in the picture.
- List the nonliving things found in the picture.

- List the dead things in the picture.
- Use words and drawings to explain how energy is flowing in the picture.
- Describe in words and pictures other ways (besides energy flow) that these living, nonliving, and dead things could be involved with each other.

Based on student responses for one class, 21 out of 26 students interchanged the terms "nonliving" and "dead" when prompted to identify these items from the woodland sketch. Plants and their parts were frequently identified as nonliving. I was very surprised by the results because I knew the students had received extensive instruction on ecology in the sixth grade.

The students' misconceptions were later confirmed in the *Benchmarks for Science Literacy* (AAAS, 1993) as common among elementary and middle school students. The research literature indicated the inability of students to define living, nonliving, and dead. I now include the identification of living, nonliving, and dead organisms throughout my thematic set, as well as initiating the unit with the abiotic and biotic factors activity. The science education research literature also provided information on children's understandings about abiotic and biotic factors. Leach, Driver, Scott, and Wood-Robinson (1992) found that children recognize that plants need soil, water, and sunlight to survive; few children, however, identified carbon dioxide or oxygen as essential abiotic factors that plants need to survive. Many children believe that plants make food for animals instead of for themselves (Roth & Anderson, 1985). Lastly, Eisen and Stavy (1988) noted that most students realize that plants produce oxygen through photosynthesis, but only about half realize that animals, because of their need for oxygen, cannot live without plants.

USING THE ASSESSMENT TASK

I initiate the biotic-abiotic pedagogical-assessment task with a pre-activity class discussion of the task, to informally assess the students' prior knowledge. I review material and introduce any new material needed for completing the assessment task. Before starting, I distribute the student response form (Figure 1). I show the students the four jars containing dilute bromothymol blue solution. Jar A contains only the bromothymol blue solution. Jar B has a sprig of elodea in the bromothymol blue; Jar C contains a snail in solution, and Jar D contains a snail and a sprig of elodea in the solution. I put lids on all the jars before I place them near the window, where they are exposed to sunlight.

The students are to note any changes taking place in the jars over a four-day period and to explain the cause of these changes using words and drawings. They are to organize and be neat in presenting their observations and explanations and complete in explaining any changes they observe between the living organisms and the environment. I place the scoring rubric, showing my expectations, on the chalkboard and I explain how I will score the response forms. The scoring rubric remains on the chalkboard throughout the activity. Because this assessment task takes place at the start of the school year, I emphasize observation and recording skills. I want to establish solid science process skills that will be used throughout the academic year. Because I want students to develop good scientific work habits, I reflect these in the scoring rubric for the activity. The response form is open-ended, so that students can record what they observe--not what they think I want them to observe. Originally I had provided students a blank sheet of paper on which they organized their observations, interpretations, and explanations in any manner they wanted. I found that this was too time consuming to assess and that I was biased toward the neatly organized papers.

DATA TABLE

HYPOTHESIS:

USE WORDS/PICTURES TO NOTE ANY CHANGES IN THE JARS OF BTB:

Day 1	Day 2
Day 3	Day 4

What caused this to happen? Be specific and as complete as possible. Use words and/or drawings for your explanation:

Figure 1. Student Response Form

Before the students begin their observations, I conduct a short demonstration with the bromothymol blue to focus their attention on the blue color of the solution. A student volunteer uses a straw and carefully blows bubbles into a test tube of the dilute bromothymol blue solution. The other students watch the solution change from medium blue to green and then yellow. Through this demonstration, students conclude that the carbon dioxide exhaled into the solution caused the color changes. Next, students predict what they think will happen if the open test tube is exposed to the air overnight. I mark that test tube and place it in the rack for observation the following day. This demonstration is necessary because students cannot discover why the bromothymol blue changes color from the in-class activity alone. Students have to understand, have prior knowledge about, why the bromothymol blue changes color. Without the knowledge that carbon dioxide causes the bromothymol blue to change color, they would fail the assessment.

After the bromothymol blue demonstration, students make a prediction and hypothesize about which color changes will occur in the jars at the end of four days. Several students then share what they have written. For the next four days, students record their observations about the color changes in the four jars compared to the open test tube. Volunteers share their observations with the class. Students organize their observations neatly on the response form, making complete observations and numbering and drawing the jars in the same sequence each day. They may use colored pencils or labels to record their observations. During the task, I check students' understandings about living, nonliving, dead, the needs of living things, and the relationship between oxygen and carbon dioxide (respiration and photosynthesis). I call on volunteers to share their observations and understandings. Based on the outcome of the class discussion, I may require students to read a section of the textbook, complete a written worksheet to review the information, or begin the assessment task. At the end of the four-day period, students review their observations and draw conclusions to explain the color changes in the jars. They check the scoring rubric to be sure they are completing all of the requirements. When the response forms are scored and returned, I guide students in connecting the bromothymol blue results to the bottle ecosystem that they will be constructing as the summative assessment for this thematic set.

SCORING THE ASSESSMENT TASK WITH AN ANALYTIC RUBRIC

One of the most important points of this assessment task is the observation and description of color changes in the bromothymol blue jars. The greater the amount of carbon dioxide in the dilute bromothymol blue solution, the more yellow the solution will become. The balanced environment includes the snail and the plant in solution is green. I expect the students to accurately observe the colors and to note any change in the organisms in the solution. I require an explanation of the observations in all of the jars each day. I developed an analytic scoring rubric that allows me to include several skills in one assessment (i.e., a multitrait assessment). A well-constructed scoring rubric displays a range of student abilities, differentiating student responses from the most excellent to the least acceptable.

The scoring rubric must be easy to follow and interpret and should align with instruction. The assessment task itself must be inexpensive to administer.

As I was developing my first scoring rubric, I felt the drawing was as important as the explanation of the assessment task; this means that I should assess the students' drawings. Drawing then became a dimension of my scoring rubric. The drawing criteria are closely related to the explanation and observation dimensions of the scoring rubric. The analytic scoring rubric illustrated in Table 1 was used to assess the multiple traits in the assessment task. In the next section, I present several samples of student work in order to illustrate the types of responses elicited by this assessment task.

Table 1. The Analytic Scoring Rubric for the Abiotic-Biotic Task

Performance Dimension And Criteria	Scoring Scheme
DRAWING (5 points total) Four jars present for each day, shown in sequence; content of jars shown; color of the solutions illustrated and neatly done.	1 point for each day that the jars drawn reflect the criteria.
EXPLANATION (5 points total) Each jar accurately explained for each day, supported with evidence from observations.	1 point for each day that all jars are accurately explained, reflecting the criteria.
OBSERVATIONS (4 points total) Observations include an accurate description of the solution's color for each jar; all (4) jars are noted for all days; the contents of each jar are presented.	1 point for each jar that is accurately observed each day, reflecting the criteria.
ORGANIZATION (3 points total) A hypothesis is stated; preplanning is shown in balancing the drawings and observations on the page; jars are drawn in same order each day; neatly composed.	1 point for hypothesis, 1 point for preplanning, and 1 point for neatness.

Student Examples

The student sample shown in Figure 2 reflects a top performance level; Jane scored 15 points of the possible 17. Her work is organized with complete drawings depicting accurate observations for the four days in all four jars. There is evidence of preplanning and understanding in the drawings and explanations. She did lose one point for the presentation of Day Two jars compared with the other days. Here she observes the green color, but does not explain what caused the solution to turn green; she provides no explanation of the plant affecting the bromothymol blue solution.

DATA TABLE

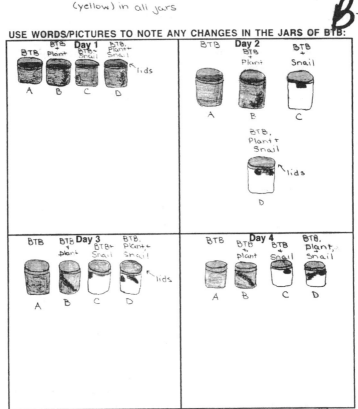

HYPOTHESIS: I predict that the BTB will turn different colors (yellow) in all jars

𝐵 -1

USE WORDS/PICTURES TO NOTE ANY CHANGES IN THE JARS OF BTB:

What caused this to happen?   Be specific and as complete as possible.
Use words and/or drawings for your explanation: Carbon dioxide causes the BTB to change to yellow and snails exhale carbon dioxide which causes the two with snails to change to yellowish-greenish color.

Figure 2. Jane's Work Sample

Bob's work demonstrates good observation and organizational skills, but his explanation for the color change to green is inaccurate (Figure 3). He states that the jar containing only the elodea turned yellow because the plant gave it extra oxygen, but he displays Jar two as blue in color. He notes that the jar containing only the snail turned yellow because of the oxygen given off by the snail. Bob received a score of 14 points.

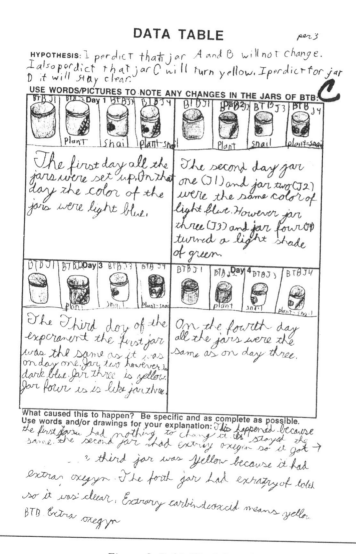

DATA TABLE

Figure 3. Bob's Work Sample

DATA TABLE

HYPOTHESIS: *The BTB will turn yellow. The plants and snail will still be alive*

USE WORDS/PICTURES TO NOTE ANY CHANGES IN THE JARS OF BTB

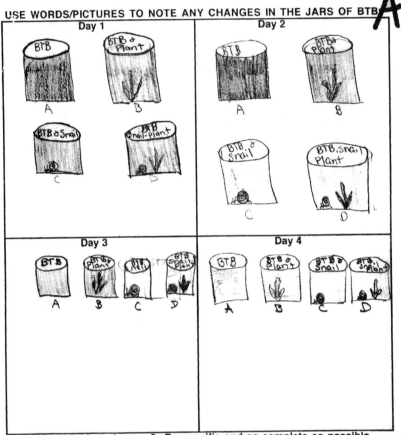

**What caused this to happen? Be specific and as complete as possible.
Use words and/or drawings for your explanation:** *The snail changed color because of waste and sunlight.*

Figure 4. Jim's Work Sample

Next, Jim's work example displays poor organization (Figure 4). The jars are not in the same sequence; the colors shown for Day Two as compared to Days Three and Four are different shades of blue, and there is no explanation to clarify his use of different colors. Further, his explanation for the changes in the bromothymol blue is incorrect: "The snail changed color because of waste and sunlight." Jim received a score of 12 points.

Finally, David's work sample reflects a low performance; it received a score of 9 points (Figure 5). David's hypothesis is not complete; all days are not represented; Day Four observations are clear in color, and the explanation is incorrect. He said that, "The snail absorbs the water and breathes it in so it becomes light blue."

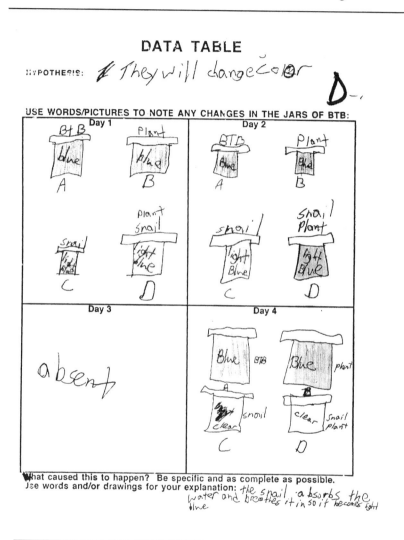

Figure 5. David's Work Sample

Several students provided no explanations.   When I interviewed them, I found they had simply forgotten to do this part of the activity in spite of the many reminders I had provided. I did not understand how this could be until I read the "Human Organisms" chapter from the *Benchmarks for Science Literacy*: "Attending closely to any one input of information usually reduces the ability to attend to others at the same time" (AAAS, 1993, p.). In the future, I will simplify this activity by making conclusions more important than practicing the skills of drawing.   One solution may be to draw containers on the student response forms rather than requiring the students to draw and organize these observations.

Misconceptions also show up in student explanations. I make an extra effort to address these through additional investigations and discussion of the bromothymol blue assessment task.   Some student misconceptions identified by the assessment task are the following:

- Heat from the plant and snail causes the color to change.
- The snail gives off oxygen, causing the color change.
- The passing of time causes the color to change.
- Sunlight causes the color to change.
- Water changes the color of the bromothymol blue.
- Chlorophyll escapes from the plant, causing the color to turn green.

These student responses provide insight into students' understandings; this informs my instructional decisions, my interaction with students and my modification of this activity. In this way, my assessment practice is aligned with NRC (1996) Teaching Standard C that encourages teachers to engage in ongoing assessment of their teaching and of student learning.

SCORING THE ASSESSMENT TASK WITH A HOLISTIC RUBRIC

After I scored one class, I felt the analytic scoring rubric was too complex and too time consuming for use with one assessment task. Students would not receive feedback in time to re-evaluate their understandings before the next activity. I was also concerned about the time I would spend scoring five class sets. I was uncomfortable with the general format of a holistic scoring rubric; however, I decided to construct one for this task. I wanted to see if it would reduce the time needed to score the students' responses as my colleagues had said it would. The holistic scoring rubric shown in Table 2 was used to assess student responses in this pedagogical-assessment activity.

Comparison of Student Scores

Because the holistic rubric listed the categories for scoring student work in performance levels, there were fewer total points. I was able to return students' work more quickly, review it with them, and take time for student reflection before the next activity in the thematic set.

Jane scored three points on the holistic rubric and 15 points on the analytic scoring rubric (see Figure 2). Examination of Bob's response (see Figure 3) shows skills of good observation; however, the explanation is inaccurate (i.e., that the extra oxygen caused the yellow color). Bob went on to say that the last jar was clear because it had extra oxygen. He scored 14 points using the analytic scoring rubric and 3 points using the holistic scoring rubric. Jim's explanation and preplanning was lacking, so I scored this response a two out of the possible four points (see Figure 4). David would also receive two points because of the range of skill abilities described in the rubric (see Figure 5).

Table 2. The Holistic Scoring Rubric for the Abiotic-Biotic Task

Performance Level	Performance Description
Excellent (4 pts.)	Contains accurate and detailed observations; contains an accurate interpretation and explanation supported with evidence; is well organized with jars labeled and appropriately colored; presentation is neat and easily interpreted.
Good (3 pts.)	Contains all observations, but is lacking in either accuracy or detail; contains a mostly accurate interpretation and explanation supported with evidence, is mostly organized with jars labeled and appropriately colored; presentation is neat and easily interpreted.
Average (2 pts.)	Some observations are missing or are incomplete; accuracy and detail are lacking; interpretation and explanation are mostly accurate but are not supported with evidence; organization is poor and jars are not all labeled or appropriately colored; messy or incomplete; some difficulties in interpretation.
Needs Improvement (1 pt.)	Incomplete observations with little accuracy or detail; interpretation and explanation are weak or without evidence; organization is poor and jars are not labeled or labeled inaccurately and are not appropriately colored; very messy or difficult to interpret; incomplete task.
Not Attempted (0 pts.)	No observations, interpretation, or explanation; no identification of jars or appropriate coloring; no attempt at communication.

The holistic scoring rubric rating of zero is only given for no response. However, a student may receive no points on the analytic scoring rubric if the responses do not match the specific criteria listed, even though that student may have responded in part. The student who responds vaguely or who possesses limited background scores better using the holistic scoring system, as demonstrated by David's score.

IMPROVING THE ASSESSMENT TASK

I will revise this assessment task again because of the number of students who provided no explanation for the color changes. I may provide students with written instructions for the assessment task; in the past these were provided orally. I plan to reduce the number of steps needed to complete this task as well as the number of verbal instructions. First, I will draw empty jars on the student response form, reducing the number of steps the students must perform to complete the assessment task. I want to see if this makes scoring easier for me and removes the bias I feel when looking at well-drawn student responses compared to ones with jars that are drawn quickly, carelessly, or in different sizes and shapes. The students would then be assessed on their observations of the color changes rather than on the quality of their presentation.

Next, I will emphasize observing more than drawing and I will score observing at a higher level. I also plan to change the heading "Explanation" to "Conclusion" on the rubric so that it is in line with the scientific format students will use to complete future activities. Revision is an important and necessary process in developing and using assessment tasks. The validity and reliability of the assessment task increases with each revision. Because each group of students reacts to the assessment task in a different way revision of the task makes it user friendly.

The first time I used this pedagogical-assessment activity, I allowed students to freely interpret the colors; another time, I told them to use only primary color descriptors. The students still questioned my description of colors that looked similar (i.e., bluish-green, greenish-blue, aqua, chartreuse). Now, I demonstrate color descriptions based on articles found in the room, asking volunteers to describe the color. While doing this, I have heard several students interchange white and clear. We now compare objects in the room that are both white and clear.

The *Benchmarks for Science Literacy* (AAAS, 1993) have been valuable in helping me plan and improve my methods of instruction and assessment. I found that middle school students have limits to the amount of information and the number of directions they can handle at any one time. These skills require practice to develop. The research reported in the *Benchmarks for Science Literacy* (AAAS, 1993), confirms this by stating:

> The level of skill a person can reach in any particular activity depends on innate abilities, the amount of practice, and the use appropriate learning technologies...Attending closely to any one input of information usually reduces the ability to attend to others at the same time (The Human Organism, AAAS, 1993, p. 141).

PEDAGOGICAL-ASSESSMENT ACTIVITIES AS SUMMATIVE ASSESSMENTS

In my thematic set on "Energy in a Living System," I use a summative assessment in which students work in groups. They are responsible for designing an eco-column investigation and observing it for five days over a two-week period.

Students are to record their investigation design and observations in their science journals. In this way, they are to apply their understandings about energy flow in a community, as well as their knowledge about the role of abiotic and biotic factors in the eco-column. Students also create a food chain using the organisms in their eco-columns. This task assesses students' science content knowledge, and their process and inquiry skills, as well as providing them the opportunity to learn as they complete the task. The scoring guide for the assessment task is displayed in Table 3.

The eco-column assessment task allows students to formulate their own investigation, enhancing the pedagogical value of the assessment activity and providing a better measure of understanding (Haertel, 1991). The eco-column assessment task, like the abiotic-biotic task, is integrated into the curriculum and provides students the opportunity to synthesize their knowledge and to develop deeper understandings (Baron, 1991).

Table 3. Scoring Guide for Eco-Column Task

Assessment Category	Rating/Score
Rationale for abiotic and biotic factors included in the eco-column is valid and clearly stated.	
Eco-column contains producers, consumers, and decomposers.	
An adequate energy source is provided for each organism.	
Energy source for each organism is correctly identified.	
The role of each organism is correctly identified.	
A valid food chain for the eco-column is stated.	
The investigation is thoroughly developed with a clearly stated question, design, and procedure.	
The investigation plan (procedure) was followed and the group was prepared for the construction of the eco-column.	
Accurate and detailed observations are recorded in the students' science journals.	
The final product is complete, with acceptable explanation of results based on evidence.	

Scoring Scheme: 3 pts. Good, no difficulties; complete and accurate
                2 pts. Average, some difficulties, incompleteness, or inaccuracies
                1 pt.   Needs improvement, many difficulties and errors
                0 pts. Not attempted or missing

STANDARDS-BASED ASSESSMENT

What constitutes standards-based assessment? Do the pedagogical-assessment activities shared in this chapter reflect standards-based assessments? The *Benchmarks for Science Literacy* broadly describes standards as, ". . . something against which other things can be compared for the purpose of determining accuracy, estimating quantity, or judging quality" (AAAS, 1993, p. 322). The *National Science Education Standards* defines standards as "criteria to judge quality

. . . criteria to judge progress toward a national vision of learning and teaching science . . . " (NRC, 1996, p. 12). The NRC (1996, p. 5) assessment standards also provide criteria for judging the quality of assessment practice:

- The consistency of assessments with the decisions they are designed for.
- The assessment of both achievement and opportunity to learn science.
- The match between the technical quality of the data collected and the consequences of the actions taken on the basis of those data.
- The fairness of assessment practices.
- The soundness of inferences made from assessments about student achievement and opportunity to learn.

Based on the above views, I believe that the assessment tasks shared in this chapter reflect standards-based assessments. I draw this conclusion because the science content and processes assessed are based on the *Benchmarks for Science Literacy* (AAAS, 1993); therefore, the content of the assessment tasks is standards-based. Further, the purpose, design, and implementation of the assessment tasks and the use of student results reflect the intent of the NRC (1996) standards. Thus, the assessment tasks themselves are standards-based.   For assessment tasks to be standards-based they must reflect both the science content standards (either the *Benchmarks for Science Literacy* or the *National Science Education Standards* or state and district standards) and the assessment standards.

CLOSING THOUGHTS ON THE VALUE OF PEDAGOGICAL-ASSESSMENT ACTIVITIES

Pedagogical-assessment activities provide students with the opportunity to demonstrate what they know and understand about science content and processes, as well as provide students with opportunities to learn science. The pedagogical-assessment activities shared in this chapter allow students to demonstrate and learn the skills of observation, organization, interpretation, record keeping, and communication.

Assessment reflects the curriculum and the methods of instruction, and is an integral, ongoing part of the classroom. The assessment task can diagnose student misconceptions and weaknesses in the students' background that need additional attention. Pedagogical-assessment activities can both evaluate student performance and diagnose student difficulties, thus informing pedagogy. The task prompt that most frequently leads to quality performance in the science classroom is open-ended and allows for free responses from students. The prompt encourages and guides students' thinking about their response and learning. These personal reflections are not possible with a multiple-choice assessment. Meaningful learning will not occur without these reflections and reconstruction of knowledge in the student's mind.

The scoring rubric defines the quality of performance expected from students. The scoring rubric may be general and holistic or it may be analytic. It provides feedback to students about how they are doing and how they can improve. Further, analytic and holistic scoring rubrics can be used to assess the science curriculum or the instructional methods. By providing insight into students' understandings and abilities, rubrics inform instruction and highlight any need for curricular revision. Each assessment tool can be refined based on the students' understandings about the world around them.

I review, reword, or revise each assessment task and scoring rubric after I reflect on the results of the assessment task. Reflection and revision are important components for improving science education. All of this takes time, but I find using alternative assessments and scoring rubrics exciting and worthwhile. I have seen students become thoughtful and more thorough in their responses. I have seen growth in student confidence while they are observing and communicating. My students reach more accurate conclusions. Pedagogical-assessment activities are one means to reform and improve science education. They are the focal points for my personal growth and continued learning. They are the "value" in the evaluation of my science program.

REFERENCES

American Association for the Advancement of Science (1990). *Science for all Americans*. New York: Oxford University Press.

American Association for the Advancement of Science (1993). *Benchmarks for Science Literacy*. New York: Oxford University Press.

Baron, J.B. (1991). Performance assessment: Blurring the edges of assessment, curriculum, and instruction. In G. Kulm & S.M. Malcom (Eds.), *Science assessment in the service of reform* (pp. 247-266). Washington, DC: American Association for the Advancement of Science.

Brooks, J.G. & Brooks, M.G. (1993). *In search of understanding: The case for constructivist classrooms*. Alexandria, VA: Association for Supervision and Curriculum Development.

Burke, K. (1992). Testing and thoughtfulness. In K. Burke (Ed.), *Authentic assessment: A collection* (pp. 49-51). Palatine, IL: IRI/Skylight Publishing, Inc.

Eisen, Y. & Stavy, R. (1988). Students' understanding of photosynthesis. *The American Biology Teacher*, 50(4), 208-12.

Haertel, E.H. (1991). Form and function in assessing science education. In G. Kulm & S.M. Malcom (Eds.), *Science assessment in the service of reform* (pp.231-245). Washington, DC: American Association for the Advancement of Science.

Leach, J., Driver, R., Scott, P. & Wood-Robinson, C. (1992). Progression in conceptual understanding of ecological concepts by pupils age 5-16. Leeds, England: Centre for Studies in Science and Mathematics Education, University of Leeds

National Research Council (1996). *National science education standards*. Washington, DC: National Academy Press.

Roth, K.J & Anderson, C.W. (1985). *The power plant: Teachers' guide*. East Lansing, MI: Institute for Research on Teaching, Michigan State University.

VICKY JACKSON

11. THE MULTIDIMENSIONAL ASSESSMENT OF STUDENT PERFORMANCE IN MIDDLE SCHOOL SCIENCE

The first thing to consider when developing an alternative assessment instrument is the purpose and the goals for assessing students; that is, why are you assessing and what is it you wish to assess about students' science learning? The purpose for the assessment instrument shared in this chapter was to measure students' conceptual understandings and performance capabilities at the conclusion of the seventh grade; in other words to serve as a comprehensive examination. The goals for the comprehensive examination also included informing practice. The *Benchmarks for Science Literacy* [*Benchmarks*] (American Association for the Advancement of Science [AAAS], 1993) assisted me as a tool in this process.

After I had determined the purpose and goals for assessment, I needed to identify the assessment format. I considered our science curriculum and the pedagogical methods employed; for example, the assessment format needed to align with both the curricular goals and instructional methods of the classroom. Just as I believe that it is unfair to teach students using laboratories and then only assess them through multiple-choice tests, it is also unfair to assess students using formats that they have not instructionally experienced. Assessment, instruction, and the curriculum must all be aligned.

THE *BENCHMARKS* AS A TOOL FOR PLANNING AND DEVELOPING ASSESSMENT TASKS

I used the *Benchmarks* as a framework for constructing the individual assessment tasks that comprised the comprehensive examination administered at the end of the seventh grade. To develop the framework I first correlated the *Benchmarks* to our seventh grade science curriculum to determine whether our program was aligned with the national standards. I also wanted to identify what our students should know about science and be able to do. I then pulled from the *Benchmarks* the content themes and the criteria for minimum competency or performance in order to determine what should be assessed. The *Benchmarks* became the classroom standards. The *Benchmarks* served as the tool for organizing my thinking about the science curriculum and what should be assessed.

181

Daniel P. Shepardson (ed.), Assessment in Science, 181—196.
© 2001 *Kluwer Academic Publishers. Printed in the Netherlands.*

Example Assessment Tasks

The assessment instrument contained a variety of assessment tasks in different formats, such that it would reach students of all levels and styles of learning. Gardner (1983) suggests that students make meaning in multiple ways based on different styles of learning. Assessment in science, too, should engage students in a variety of tasks or experiences that provide students with different avenues for demonstrating their science learning. Teachers need to provide situations, including assessments that engage students in multiple ways of thinking and representing knowledge (Adams & Hamm, 1998). If a student does not perform well on one assessment task format, that student may have more success with a different assessment task format. By administering multiple assessment formats to students, students have multiple opportunities to demonstrate their science learning; I am better informed about their understandings and capabilities. This, in turn, helps me to design more effective instruction.

After identifying and developing the assessment task formats, I utilized the *Benchmarks* to assist me in writing the scoring rubrics for each task. For example, the three benchmarks below provided the standards for a scoring rubric where students were to interpret a population graph depicting a relationship between deer and wolves. The task required students to demonstrate an understanding of predator-prey relationships and the ability to read and interpret simple graphs:

> Two types of organisms may interact with one another in several ways: They may be in a producer/consumer, predator/prey, or parasite/host relationship. Or one organism may scavenge or decompose another. Relationships may be competitive or mutually beneficial. Some species have become so adapted to each other that neither could survive without the other. (The Living Environment, 5D Interdependence of Life, AAAS, 1993, p. 117).

> Read simple tables and graphs produced by others and describes in words what they show. (Habits of Mind, 12D Communication Skills, AAAS, 1993 p. 297).

> Physical and biological systems tend to change until they become stable and they remain that way unless their surroundings change. (Common Themes, 11C Constancy and Change, AAAS, 1993, p. 274).

Based upon the above benchmarks I constructed the following top-level performance rubric for assessing students' understandings and abilities:

Level 3 (Top Performance). The deer population increased because the predator population decreased from 350 to 40. The predators (wolves) are those animals that eat the deer for food, the prey. The change (decrease) in the predator population resulted in a change (increase) in the deer population.

The scoring rubric reflects the "Habits of Mind" benchmark in the criterion, "The deer population increased because the predator population decreased from 350 to 40." This criterion assesses students' ability to read graphs and to describe in words what graphs show. The scoring rubric criterion, "The predators (wolves) are those animals that eat the deer for food, the prey," assesses students' understanding of the interdependence of life. This criterion illustrates the predator-prey relationship: the interaction between wolves that eat the deer for food. The "Constancy and Change" aspect of the scoring rubric is detailed in the last sentence: "The change (decrease) in the predator population resulted in a change (increase) in the deer population." This criterion relates to the idea that biological systems (animal populations) remain stable unless their surroundings change. These benchmarks provided me with a sense of the understandings a scientifically literate student should have for each assessment task, and ensured that the scoring rubric addressed what I was trying to assess. I had to make a determination about the scoring rubric and the *Benchmarks* provided me with the criteria to make that determination.

Although the *Benchmarks* served as a guide for writing the scoring rubrics, I still encountered difficulties in scoring student responses or performances on several assessment tasks. I had to rework the scoring rubrics; in most cases the difficulty was a lack of specificity that prevented me from differentiating the level of student performance. Also, student responses reflected ideas, procedures, or interpretations that I had not thought about, but which were acceptable for accurate completion of the task. For these reasons, I revised the scoring rubric, incorporating these new insights about students' science learning. Student responses also provided insight into which assessment tasks were stated in a confusing manner. I revised several assessment tasks that were confusing or misleading based on student responses.

In the next section, I share several of the assessment tasks that represent the different formats used to assess student performance. I have selected three assessment tasks that I believe illustrate unique ways of determining students' science learning. I have included student responses and a critique of these responses to illustrate the application of the scoring rubric.

Example Assessment Task: Writing a Newspaper Article

As previously indicated, I used student responses as a resource for reflecting upon and informing my teaching. For example, in the newspaper writing task (Figure 1), no student scored at the top performance level. In this situation, I felt that I had not provided students with sufficient writing experiences throughout the academic year so that they could successfully complete the task. I need to provide my students with more practice and feedback on completing tasks of this kind throughout the academic year. I also need to work with other faculty team members so that there is an interdisciplinary aspect to writing responses of this type, and so that students know what the expectations are. Students will be better prepared if they complete similar assessment tasks in mathematics, social studies, and English.

Imagine that you are a science writer for a newspaper. Write a newspaper-style article for one of the following headlines. Be sure to give your readers scientific information about the headline.

a) Amateur Astronomer Discovers New Comet
b) Supernova discovered in Andromeda Galaxy
c) Near Collision Between Asteroid and Space Probe

Scoring:

Exemplary response (5 points)
Responses are correct and supported with complete details of the scientific knowledge associated with headlines.

a) Includes description of what an astronomer is and information about comets, what they are made of, and where they are found. Students may also give history of previous comets discovered.
b) Includes information about what a Supernova is and where it fits into scheme of star life cycle. Answer would also describe galaxies.
c) Includes facts about asteroids and where they are found. Answer would also discuss use of probes to discover more about the solar system. Answer should address questions about why probe would be near the asteroid.

Proficient response (4 points)
Responses are correct but lack details of scientific knowledge associated with headline.

Adequate response (3 points)
Most responses are correct and are supported with some details of scientific knowledge associated with headline.

Inadequate response (2 points)
Many responses are incorrect and/or lack details.

Incomplete response (1 point)
Begins, but fails to complete tasks.

No response (0 points)
No attempt to answer question.

Figure 1. Prompt and Scoring Rubric for Writing a Newspaper Article

Although no students received a Level 5 score, several students scored at a Level 4 (Figure 2). The student example shown in Figure 2 reflects a Level 4 score because while the student had a good grasp of the content, he had difficulty in thoroughly presenting the facts and details about discovering a new comet. For example, the student response includes information about comets ("Usually get smaller as they pass the sun"), what they are made of ("Consisting of ice"), and for the most part where they are found ("Path takes it farther away from earth"). A stronger description of where comets are found, along with a description of what an astronomer is, would be necessary to score at a Level 5. The *Benchmarks* describe an understanding of comets as follows:

> Other chunks of rocks mixed with ice have long, off-centered orbits that carry them close to the sun, where the sun's radiation (of light and particles) boils off frozen material from their surfaces and pushes it into a long, illuminated tail (The Physical Setting, 4A The Universe, AAAS, 1993, p.64).

Imagine that you are a science writer for a newspaper. Write a newspaper-style article for one of the following headlines. Be sure to give your readers scientific information about the headline.

d) Amateur Astronomer Discovers New Comet
e) Supernova discovered in Andromeda Galaxy
f) Near Collision Between Asteroid and Space Probe

An amateur astronomer, Joe Smith, discovered a comet today. Comets consisting of ice and dirt usually get smaller as they pass the sun, however this comet was fairly large, with its tail being 1.7 million miles long. It is assumed that it is a fairly new comet. Its path takes it farther away from Earth then Halley's comet so an educated guess was made that it would be back in 112 to 115 years. Mr. Smith, having just recently bought a telescope, was trying to focus on what he thought was the moon but he figured that it was too small and that "the thing kept moving".

Figure 2. Level 4 Student Work Example for Writing a Newspaper Article

Example Assessment Task: Drawing and Explaining

The next assessment task deals with constancy and change and requires students to draw a series of pictures and write an explanation about succession (Figure 3). The series of drawings should align with and give meaning to the written explanation; conversely, the written explanation should align with and give meaning to the drawings. The drawings require students to construct a picture of their mental representations, providing another avenue for determining their understanding of succession. Students who have difficulty with verbal or written explanations, but who are more visual, can express their understandings through drawings. Some students have the ability to use language (verbal-linguistic) to convey meaning while others may be more spatial-visual in their ability to convey their understandings (Gardner, 1983). The ultimate aim is for students to convey their understandings through both writing and drawing.

Based on the *Benchmarks*, the standard for this task would reflect that ". . . biological systems tend to change until they become stable and then remain that way unless their surroundings change" (Constancy and Change, AAAS, 1993, p. 274). Using this benchmark to guide my scoring rubric development, the following holistic scoring rubric was constructed to assess student responses:

Top Level Performance (Level 3): Drawings clearly illustrate a change in vegetation, ranging from barren ground to herbaceous plants or shrubs to woody plants. Drawing is aligned with explanation. Explanation is complete, indicating that vegetation changes over time from barren ground to herbaceous plants or shrubs to woody plants, and that the vegetation will change until it becomes stable and then remain that way unless there is a change or disturbance. Identification of specific plants at each successional stage is required.

Acceptable Level of Performance (Level 2): Drawings clearly illustrate a change in vegetation, ranging from barren ground to herbaceous plants or shrubs to woody plants. Drawing reflects explanation. Explanation lacks some details in indicating that the vegetation changes over time from barren ground to herbaceous plants or shrubs to woody plants; and that the vegetation will change until it becomes stable and then remain that way unless there is a change or disturbance. No identification of specific plants at each successional stage.

Unacceptable Level of Performance (Level 1): Drawings and explanation are incomplete, incorrect and/or not clear.

Not Attempted (Level 0): No response provided.

The above scoring rubric is a revision from my first scoring rubric, which lacked specificity in describing student responses and which did not clearly illustrate how the benchmark for constancy and change was integrated into the scoring rubric. The first scoring rubric follows: .

- 4 points - Student drawing correctly identifies possible sequence of events and explains with complete details.
- 3 points - Student drawing correctly identifies possible sequence of events but details are incomplete.
- 2 points - Student drawing correctly identifies possible sequence of events but explanation is missing.
- 1 point - Drawing or explanation is incorrect and/or incomplete. Student attempts to answer but fails task.
- 0 points - No attempt is made to answer question.

The year is 2010 and you are preparing a garden. The land is plowed, raked and ready to go, but something happens and the garden never gets planted. In the boxes below, draw what you think the garden will look like at the end of the summer, after one year, and after several years. Explain why you think this is so.

End of Summer

One Year Later

Several Years Later

Explain what is happening in your pictures and why.

Figure 3. Prompt and Student Response Form for Drawing and Explaining

Table 1. Scoring Rubric for Drawing and Explaining Task

| Performance Level | Performance Categories | | |
	Drawing Clarity	Drawing-Explanation Alignment	Explanation
Top Level	Drawings clearly illustrate a change in vegetation, ranging from barren ground to herbaceous plants or shrubs to woody plants.	Explanation aligns with drawings.	Explanation is complete, indicating that vegetation changes over time from barren ground to herbaceous plants or shrubs to woody plants, and that the vegetation will change until it becomes stable; it remains that way unless there is a change or disturbance. Identification of specific plants at each successional stage is included.
Acceptable Level	Drawings do not clearly illustrate a change in vegetation, ranging from barren ground to herbaceous plants or shrubs to woody plants.	Explanation partially aligns with drawings.	Explanation lacks some details about the change in vegetation over time from barren ground to herbaceous plants or shrubs to woody plants, and the idea that the vegetation will change until it becomes stable; it will remain that way unless there is a change or disturbance. No identification of specific plants at each successional stage is included.
Unacceptable Level	Drawings inaccurately illustrate a change in vegetation.	Drawings do not align with explanation.	Explanation is inaccurate.
No Attempt	Drawings are missing.	Drawings and explanations are missing.	Explanation is missing.

Although it was clear to me what "explained with complete details" meant, it was not clear to other teachers or students. The general scoring rubric was easy to write, but was limited in its use because it lacked specificity. In addition, I collapsed two of the levels based on student responses. The revised scoring rubric, because of its specificity, ensures better consistency in scoring student responses. I considered constructing the scoring rubric in a more analytical manner, providing a score for drawing clarity, for alignment between drawings and explanation, and for the explanation, but I decided against this practical reasons. Had I revised the scoring rubric to reflect three performance categories (drawing, alignment, and explanation), it would have been similar to the rubric presented in Table 1.

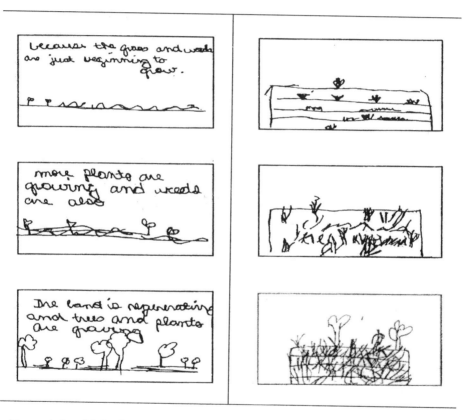

Figure 4. Level 3 Student Work Example for Drawing and Explaining   *Figure 5. Level 2 Student Work Example for Drawing and Explaining*

The two student work examples illustrate a Level 3 (Figure 4) and a Level 2 (Figure 5) response. Although the Level 3 student includes both drawings and written explanations, the drawings do not clearly illustrate succession. The written explanation is also vague, especially for the last two stages, "One year later" and "Several years later." The student also failed to explain that the system becomes stable and remains stable unless the surroundings change or are disturbed. I would expect a top-level response to provide specific examples of vegetational change, to contain clearer drawings, and to describe the constancy and change aspects of succession. On the other hand, although the Level 2 student fails to provide a written explanation, the drawing conveys a vegetational change over time, from partially barren ground for "end of summer" to thicker, bushier vegetation for "several years later."

Example Assessment Task: Planning and Conducting a Laboratory Investigation

Planning and conducting a laboratory investigation is a practical task that assesses multiple dimensions of student performance (Figure 6). I wanted to know what my students understood about planning and conducting laboratory investigations using Vee diagrams. Vee diagrams (Novak & Gowin, 1984) are an integral component of instruction at the seventh grade, so the students had numerous opportunities to work with them. Specifically, I wanted to find out if my students understood experimental design, if they could use laboratory equipment and materials, and if they could carry out the laboratory experiment they had designed. The organization of data was also important and was one of the benchmarks for which I wanted to assess students' abilities: could they present their data in an organized fashion and then use the data to draw a conclusion? This task assesses student performance on four traits: 1) use of equipment and materials, 2) experimental design and Vee diagrams, 3) organization of data, and 4) ability to form a logical conclusion based on observations and evidence (see Figure 7 for example scoring sheet). For each of these traits a scoring rubric was developed to assess student performance (see Appendix A).

The benchmarks for "Scientific Inquiry" and "Communication Skills" guided the development of this assessment task and scoring rubric. The benchmarks for scientific inquiry (AAAS, 1993, p. 13) state that student should know that:

> Although there is no fixed set of steps that all scientists follow, scientific investigations usually involve the collection of relevant evidence, the use of logical reasoning, and the application of imagination in devising hypotheses and explanations to make sense of the collected evidence (Scientific Inquiry, AAAS, 1993, p. 13).

> If more than one variable changes at the same time in an experiment, the outcome of the experiment may not be clearly attributable to any one of the variables (Scientific Inquiry, AAAS, 1993, p. 13).

The benchmark for scientific inquiry provided insight into the experimental design and Vee diagram component of the task, where students design an investigation and identify and control variables. The component dealing with student ability to draw logical conclusions from observations and data is also supported by the benchmark for "Scientific Inquiry." The benchmarks for communication skills state that students should be able to:

Write instructions that others can follow in carrying out a procedure. Use numerical data in describing and comparing objects and events (Communication Skills, AAAS, 1993, p. 296).

Organize information in simple tables and graphs and identify relationships they reveal (Communication Skills, AAAS, 1993, p. 297).

The experimental design and Vee diagram component also align with the communication skills benchmark because students need to write clear instructions about how they are going to conduct their laboratory investigations. The assessment component requiring students to organize data, use numerical data, and construct a data table is also based on the benchmark for "Communication Skills."

A Hard Problem

You have recently been hired by a company selling water softeners to do some water testing. Your first assignment is to find out if the amount of calcium chloride in water affects the ability of the water to make suds. Using your scientific abilities to solve this problem, design and carry out an experiment that will give your company an answer to this question. The company would like you to report to them on the Vee diagram provided.

Before you Begin. . .

As you work through your tasks, keep in mind that your supervisor (Mrs. Jackson) will be observing:

- How you use the materials and equipment
- How clear and complete your experiment is
- How well you draw conclusions from your observations and data

Task 1

Complete the Vee diagram except for the data, conclusion, and application. Make sure you clearly outline the steps of your experiment and that you make a data table for your observations.

Task 2

Conduct your experiment with the equipment and materials in the lab. Make sure you clean your station when completed.

Task 3

Once the experiment is complete, finish filling out the Vee diagram and turn it in to Mrs. Jackson.

Figure 6. Prompt for Practical Assessment Task Item

Student Names	Use of Equipment and Materials (4 pts.)	Experimental Design and Vee (5 pts.)	Organization of Data (3 pts.)	Ability to form Logical Conclusion based on Observations (3 pts.)

Figure 7. Scoring Sheet for Practical Assessment Task Item

For the practical task, students first designed their laboratory experiment using the Vee diagram. They then carried out the laboratory experiment and completed the Vee diagram by entering their data and conclusions. To assess students' use of equipment and materials, I observed them performing the laboratory experiment and then gave them a score based on the rubric. I then assessed each student's completed Vee diagram. The completed Vee diagram served as a record or evidence of student performance in designing an experiment, in organizing data, and in drawing conclusions.

The student work example shown in Figure 8 reflects a top performance level in three of the four categories. The Vee diagram is complete and the student designed the experiment in a workable fashion, describing in detail the procedures and the controlling of variables (i.e., calcium chloride). The experiment was clearly and completely written out; communicating that the student knew exactly what she was going to do and others could follow the procedures. This student scored at a Level 5 on experimental design and the use of the Vee diagram.

The student conducted the laboratory investigation using the equipment in an appropriate manner, accurately measuring the distilled water and calcium chloride, and following the laboratory safety rules, particularly wearing safety goggles. This student received a performance score of 4 for use of equipment and materials. The student's data table reflects a Level 2 performance because the data table does not indicate the units of calcium chloride or the number of drops of soapy water added to the solution. The data table does, however, contain acceptable headings and conveys a qualitative measure of the soapsuds. Thus, the table is mostly correct, but lacks some information: a Level 2 performance. The student's conclusion matched the data and answered the focus question, providing evidence to support the conclusion, "The one without $CaCl_2$ has more suds. So calcium chloride can affect the amount of suds." Further, the student's application response was consistent with what had been learned from the experiment. Therefore, this student received a Level 3 score on the ability to a draw logical conclusion based on observations. What is important, however, is not the "correct answer," as much as the ability to design an experiment, to be consistent and logical, and to clearly explain what to do with the data collected.

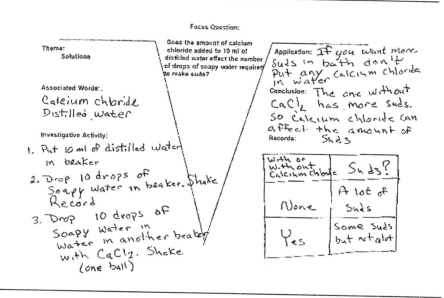

Figure 8. Student Work Example for Practical Assessment Task Item

DESIGNING AND IMPLEMENTING ALTERNATIVE ASSESSMENT: LESSONS LEARNED AND ADVICE FROM THE FIELD

Developing and using alternative assessment instruments is not easy, but it is worth the effort because it provides a more complete picture of students' abilities. Based on my experience with developing a comprehensive and alternative assessment instrument, I recommend beginning the process by first examining the curriculum and pedagogical methods used to deliver the curriculum. Then decide what is important, what you really want to assess to see what your students have learned or can do. It is important not only to find out what students have learned, but also to revise your science curriculum and teaching based on student performance. Finally, assessments need to include conceptual understanding, going beyond simple factual knowledge.

To field-test and revise my assessment tasks, I not only analyzed student responses and performances on the tasks, but I also talked to students after they had completed the assessment tasks to understand their thinking about and perceptions of the assessment tasks. I found that students did not feel as if they were being tested, especially on the practical tasks. My students indicated that they enjoyed some of the assessment tasks because they were challenging and more active then typical tests. Students also felt that there were more opportunities to think through the assessment task instead of simply trying to guess the right answer on a multiple-choice test. I think my students felt more confident than they had with a multiple-choice test, but students who are used to multiple-choice, matching, and true and false questions will need time to adapt to alternative forms of assessment.

Writing and using scoring rubrics was not easy at first. The difficulty I encountered was writing a rubric that would be fair and that other science teachers could use. It was also difficult to express my thinking and rubric criteria using specific language. The *Benchmarks* assisted me in overcoming this difficulty; I also consulted with other teachers and looked at examples from other assessment instruments for ideas about writing scoring rubrics. The next constraint that I encountered was time to develop assessment tasks and to use and score them. With practice, however, the time required to score assessment tasks is reduced. Because alternative forms of assessment give me a better picture of my students' abilities, I value the time it takes to assess them. I have resolved the time conflict by covering less science content.

I also encountered a management problem with the laboratory-based assessment component. I had an initial problem trying to watch four or five students conducting laboratory investigations at one time. To overcome this management constraint, I took advantage of the multi-dimensional nature of the assessment instrument and scheduled students at stations so that I could observe their performance while other students were completing the non-laboratory component. I learned not to try and assess every skill or ability in one task. I decided to spread performance time over a couple of days and not try to do everything at once. As science teachers, if we truly value student performance, time will not be a constraint in implementing alternative forms of assessment. On the other hand, if we do not value student performance as a means for demonstrating understandings, skills, and abilities, time will prevent the use of alternative forms of assessment in science. As teachers we have become weighed down and unfocused about what we are actually assessing. In the past, we really did not know why we were giving a test, except that the test was at the end of the chapter, or the school district required grades. In fact, the test really did not assess what we valued, what we wanted our students to know or be able to do in science.

ACKNOWLEDGEMENT

Portions of this chapter appeared in Shepardson, D.P. & Jackson V. (1997). Developing alternative assessments using the *Benchmarks*. *Science and Children*, 35(7), 34-40.

REFERENCES

Adams, D. & Hamm, M. (1998). *Collaborative inquiry in science, math, and technology*. Portsmouth, NH: Heinemann.

American Association for the Advancement of Science (1993). *Benchmarks for Science Literacy.* New York: Oxford University Press.

Gardner, H. (1983). *Frames of mind.* New York: Basic Books.

Novak, J.D. & Gowin, D.B. (1984). *Learning how to learn.* New York: Cambridge University Press.

APPENDIX A: SCORING RUBRIC FOR THE PRACTICAL ASSESSMENT TASK ITEM

Use of Equipment and Materials

4 points   All equipment is used appropriately and accurately. Safety rules are observed and goggles are worn. Lab is cleaned up after the experiment is completed.

3 points   All equipment is used appropriately and accurately and most safety rules are observed, but student must be reminded to wear goggles. Clean up after lab is complete.

Or

All equipment is used appropriately and accurately and safety rules are followed, but clean up is incomplete

2 points   Most equipment and materials are used appropriately and safety rules are observed, but student must be reminded to wear goggles; clean up is incomplete.

1 point   Student does not demonstrate knowledge of how to use equipment and/or materials. All safety rules are not followed.

0 points   Student makes no attempt to use equipment supplied.

Ability to form a logical conclusion based on observation

3 points   Conclusion is logical and clearly stated; it is strongly supported by collected data.

2 points   Conclusion is fairly logical with some data to support it.

1 point   Conclusion is not supported by collected data.

0 points   Conclusion is not stated.

Experimental Design and Vee

5 points   Design of experiment is workable and clear. Vee diagram is complete
            and is neat and shows details. Use of controls and variables is correct.

4 points   Design of experiment is workable. Vee diagram is complete but may
            lack details needed to be clear. Controls and variables are used
            correctly.

3 points   Design of experiment may be workable. Vee diagram is mostly
            complete but is not neat and lacks details. Controls and variables are
            used.

2 points   Design of experiment may not work to solve problem. Vee diagram is
            incomplete. Controls and variables may not be used or are used
            incorrectly.

1 point   Design of experiment is unclear and incomplete. Vee diagram is
            incomplete.

0 points   Student does not attempt task

Organization of data

3 points   Data table is complete and easy to interpret; it includes heading, units,
            and it is neat.

2 points   Data table is mostly correct but some information may be missing.

1 point   Data table is confusing and missing too many details. Data is largely
            organized.

0 points   No attempt has been made to organize data.

TED LEUENBERGER

12. DEVELOPING AND USING DIAGNOSTIC AND SUMMATIVE ASSESSMENTS TO DETERMINE STUDENTS' CONCEPTUAL UNDERSTANDING IN A JUNIOR HIGH SCHOOL EARTH SCIENCE CLASSROOM

This chapter outlines the process of developing worthwhile diagnostic and summative assessment tasks, as well as developing analytic and holistic scoring rubrics. The chapter provides examples of assessment tasks, scoring rubrics, and student responses as well as data that show the appropriateness of open-ended response items over selected response type tests.

Diagnostic assessments are often ignored as an integral part of instruction and summative assessments usually take on a more traditional format using selected response items such as multiple-choice, matching, or true and false. Early in my career I used pre-tests that were primarily multiple-choice and never found them to be particularly insightful. Since then I had always dismissed the importance of diagnostic assessments because I felt that after, 20 years of experience, I knew the needs of my students. I felt that any diagnostic assessment would only confirm what I already knew about my students' level of understanding. After using an alternative form of a diagnostic assessment, I found that very valuable insights are missed when diagnostic assessments are not done or done well. The most important and useful information that I have found using diagnostic assessments is that students have so many misunderstandings.

My summative assessments had always been dominated by selected response items (e.g., multiple-choice, matching, and true and false questions). I remember being convinced by my college courses that anything could be accurately tested by a selected response question. With the advent of the SCANTRON grading machine the selected response test seemed to be the only choice. When I started to experiment with alternative forms of assessment, I discovered that I was missing a multitude of insights about my students that go unnoticed on a multiple-choice test. I also found that students could more properly express themselves on an alternative assessment format. Multiple-choice tests prevented students from learning how to express themselves, a great disservice, and it reduced students' ability to think through the assessment.

As I began to use alternative assessments, I began to think about what it means to be scientifically literate. I used to think that a student who did well on a selected

Daniel P. Shepardson (ed.), Assessment in Science, 197—209.
© 2001 *Kluwer Academic Publishers. Printed in the Netherlands.*

response or constructed response (e.g., fill in the blank, short answer questions) test was scientifically literate, because the test score demonstrated that the student had gained knowledge and competencies in science. A scientific literate individual, however, is one who:

- Understands science concepts and principles
- Has a capacity for scientific thinking
- Uses science knowledge to think about and make sense of events and situations that they encounter in everyday life
- Comprehends scientific explanations (American Association for the Advancement of Science [AAAS], 1990; 1993)

Scientific literacy goes beyond knowing science facts and involves understanding and applying science concepts and principles to explain events and situations. Assessment from this perspective would require students to demonstrate conceptual understanding, use science concepts and principles to explain situations and events, and engage in thinking and reasoning about problems and situations.

I also began to think differently about what constituted student learning and achievement. I used to think that learning was reflected in an increase in achievement as measured by selected response and constructed response formats or tests. From my point of view, students who scored higher on tests showed higher achievement and learning. I now view learning as the process involved in changing or constructing understandings, knowledge, ability, and skills through experience (Wittrock, 1977). Learning is the active construction or reconstruction of cognitive structures based on social interactions and experiences. On the other hand, achievement is the numerical score one receives on an individual assessment; it reflects a snapshot of student ability, but not necessarily any change or development in understanding, knowledge, ability, or skill. For me, learning reflects change and achievement reflects application. In other words, "The performance potential acquired through learning is not the same as its reproduction or application in any particular performance situation" (Good & Brophy, 1986, p.134).

This distinction between learning and achievement has profound ramifications for both teaching and assessment (Cizek, 1997). Assessment of science *learning* emphasizes measuring change in students' understanding, knowledge, science process and inquiry skills, and even in science attitudes. The measurement of student performance on an individual assessment task more accurately reflects *achievement*. On this view, diagnostic assessments become a necessary component of classroom assessment practice. Diagnostic and summative assessments work together to determine student learning, change and development in conceptual understanding, knowledge, and science process and inquiry skills.

In contrast, traditional methods of assessment, such as multiple-choice tests, do not always identify specific needs or misunderstandings. Selected response formats provide students with the opportunity to guess at a correct response but often do not include the response that the student would actually select. Alternative assessment techniques more accurately assess students thinking and understanding, better identifying the specific needs of the students (Wolf, Bixby, Henn, & Gardner, 1991). Because students must express their ideas on an alternative form of assessment, well-defined scoring rubrics provide a more accurate measurement of understanding.

COMPARING ASSESSMENT FORMATS

To differentiate the effectiveness of assessment formats, I administered both alternative and selected response forms of diagnostic and summative assessments to a group of 88 seventh-grade students during one academic year. Both assessment formats revealed a significant increase in student understanding from the beginning of the unit to the end.   The multiple-choice test, however, showed more inconsistency in individual scores. Using the multiple-choice test results, only 17% of the students showed a significant increase in understanding (i.e., 20% or greater improvement in score). Most students, 76%, showed little change (i.e., less than 20% change in score), and 7% showed a significant decrease (i.e., 20% or greater decrease in score). This suggests that guessing and luck play a major role in determining individual scores in selected response formats (especially during the diagnostic assessment). The alternative assessment, which elicited open-ended responses, seemed a much better indicator of individual understanding. On the open-ended response format, no student showed a decline in understanding: one out of 88 showed no improvement (scoring exactly the same on both assessments), and 48% of the students showed significant improvement. The selected response format did a fine job of evaluating the achievement of the entire class, but lost value and became suspect when used as an indicator of individual understanding and learning.

In summary, an open-ended response format can reveal student understandings and misunderstandings that will not appear in selected response formats because students have a chance to express themselves in ways that more clearly indicate their conceptual level. Diagnostic assessments can reveal student misunderstandings that interfere with instruction if not identified and addressed in teaching.   These assessments can also identify students who hold scientifically accurate conceptual understandings and who need a different level of instruction.   Diagnostic assessments also prepare students for instruction by engaging them in thinking about the science concepts embedded in the assessment.

PLANNING AND DEVELOPING DIAGNOSTIC AND SUMMATIVE ASSESSMENTS

The first step in planning diagnostic and summative assessments is to identify the main conceptual focus of the science topic to be taught. Attention should also be given to the application of previously learned concepts to these new conceptual goals. Each goal should be assessed by a single open-ended response task. Assessing too many concepts in one assessment task may introduce confounding variables that confuse students, resulting in an inaccurate measure of their conceptual understanding or performance. This also makes it difficult to develop a scoring system that reliably assesses all of the relevant concepts.

When I was preparing to teach a unit on the movement of air, I identified the need to understand convection. Previous to this unit, students had learned about density and how differences in density cause objects to float or sink. Therefore, my assessment task had to show what students understood about what happens if air is heated; it should rise and this should result in a current of moving air. Students' use of the term "density" to describe the resulting motion of air also had to be assessed. I had always used a simple convection box with a candle to demonstrate convection and felt that this would make a worthwhile assessment task in an open-ended response format.

The First Attempt: Problems Encountered

The first year that I used the convection box diagram (Figure 1), I placed a real convection box in front of the class and handed the assessment to the 98 students in the class. Only two students described the correct motion and used the term "density" correctly. Most students responded with limited understanding. Several students had responses like, "The flame will run out of air and go out." This type of response confused convection with combustion. These students appeared to perceive the task from a different conceptual perspective, focusing in on the burning flame instead of air heating. I found that many students simply did not understand the convection box. This illustrates another important point about alternative assessment tasks; that it is imperative that students understand the assessment task. Clearly, if students do not understand the assessment task their responses will be adversely affected.

I also found that most students did not use complete sentences or complete ideas in the explanation. Using nonspecific words also made student responses unclear. A response such as, "It went up because of the candle," may infer that the air rose because of the heat from the candle, but the student's meaning is unclear. "It" might refer to "air" or it could refer to "density" or to something else. "The candle" may not refer to the heat from the flame. In an open-ended response assessment, inferences about the meaning of words must be monitored. It is very important that student responses be complete and that they not include ambiguous terms like "it." The use of precise language is an essential indicator of students' understandings (Costa, 1984). As students become more knowledgeable, language becomes more precise, concise, descriptive, and coherent. Some students, however, simply wrote, "I don't know." This response was also unsatisfactory because every student needs to express some idea about how the system might react by using prior ideas and experiences as sources of data to support their explanations (Costa, 1984).

Packet 3 Pre-assessment   Name_____ Group ___

Compare the diagram to the box that is on the front desk. After the explanation of how the box is constructed, draw (using arrows) how the air might be moving. Explain why you think this is true. Use the term "density" if you can.

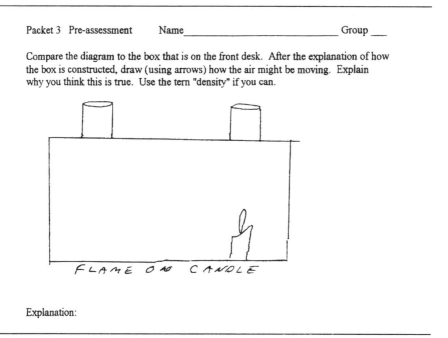

Explanation:

Figure 1. Convection Box Assessment Task

Improving the Assessment

In my first attempt I found that students needed practice in formatting their responses. Therefore, when administering an open-ended response task:

- Instruct students to use complete thoughts (sentences) avoiding ambiguous terms (e.g. "it").
- Give clear directions and provide a good explanation of the task.

With the convection box assessment, I now describe the box to the students, showing them the openings and even lighting the candle (since air is invisible). To provide incentive to students who respond with, "I don't know," I give a grade on the diagnostic assessment based on a holistic rubric that assesses the process but not the concepts expressed. I now feel that I get an accurate view of each student's level of conceptual understanding.

Scoring Diagnostic and Summative Assessments

The method used to score student responses is an essential component of successful diagnostic or summative assessments. I use an analytic rubric to assess student responses on both diagnostic and summative assessments. To create an analytic rubric, I first decide exactly what I want to find out from the open-ended response task. For example, with my convection box task, I wanted to see if students would realize the appropriate movement of air as a result of heating. I also wanted to see if they could apply the term "density" in the explanation. Therefore, I developed three analytic rubrics: one to assess the students' diagram, one to assess students' understanding of convection, and one to assess students' use of the term "density" (Table 1).

Table 1. The Analytic Rubric for Diagram, Convection, and Density

Diagram Rubric

Student showed no arrows	0 pt.
Student showed incorrect or confused motion	1 pt.
Student showed air rising over the candle but no resulting motion elsewhere	2 pts.
Student showed most motion but not complete motion	3 pts.
Student showed a convection current with air rising over the candle and air sinking down the opposite tube with air moving across the box	4 pts.

Convection Rubric

No response (including "I don't know")	0 pts.
Incorrect explanation or an observation (not an explanation)	1 pt.
Explanation includes the idea of heat or hot air rising	3 pts
Explains the existence of a current as a result of heating	4 pts.

Density Rubric

Does not use the term "density" or uses it incorrectly	0 pts.
Confuses "more dense" and "less dense"	1 pt.
Uses density, or one of its forms, correctly	2 pts.

This scoring system allows me to know exactly how each student responds. On the diagnostic assessment, I record each score separately; as a summative assessment, I combine the score with the rest of the assessment to generate a test score. On the density rubric, a score of 1 identifies a student who has the right idea but who has confused "less dense" with "more dense." The scoring rubric is designed to capture students' understandings or misunderstandings. If students express different conceptual understandings from those reflected in my scoring rubric, I note this and revise the scoring rubric to reflect these understandings, so that in the future my scoring rubric better captures students' understandings.

A holistic rubric may be included to assess the effort and form used by the students and may be applied to any open-ended response task (see Table 2 for a sample rubric). The holistic rubric may be used as a grading device, especially for diagnostic assessments, as it provides an external incentive for students who do not try to complete the diagnostic assessment because it does not count towards a grade. Knowing they will be graded on their response form ensures that students respond.

This, in turn, has a positive effect on the reliability of the diagnostic assessment to elucidate students' conceptual understandings.

Table 2. Holistic Rubric for Assessing Student Response Format

Performance Level	Score
Performance Level 2. Followed all directions, used complete sentences, provided complete thoughts, and avoided the use of nonspecific terms.	6 pts.
Performance Level 1. Had difficulty following directions, using complete sentences, providing complete thoughts, and/or avoiding the use of nonspecific terms.	3 pts.
Performance Level 0. Fails to respond.	0 pts.

Examples of Student Responses

Examples of two student responses to the diagnostic assessment task are shown in Figure 2. The first set of scores comes from Rubrics A (Diagram Rubric), B (Convection Rubric), and C (Density Rubric). These scores are not applied as grades, but are used to provide students with feedback. These same scores also provide me with information about the students' conceptual understandings. The final score is based on the holistic rubric and is recorded in the grade book.

Based on this scoring system, student number 1 (Figure 2a) received a response form rating at Performance Level 1, because he used nonspecific terms. He did follow the directions and provided a complete response. The diagnostic task tells me that this student does not understand convection, how heat effects air movement. This conclusion is based on the fact that the student incorrectly illustrates the air movement in the convection box; the arrows point toward the candle. This suggests that the candle flame uses the air. The student's explanation supports this erroneous drawing, indicating that the ". . . flame needs air to feed it so all the air would go to the flame." The student does not incorporate density in his explanation at all.

The second student (Figure 2b) received a top performance level because she followed the directions, provided complete thoughts, and avoided the use of nonspecific terms. Conceptually, this student had difficulty explaining air movement in the convection box. The student's diagram is incorrect and the explanation indicates that the air will exit the convection box on the left side because the rising smoke exits the box on the right. Finally, this student uses the term "density" incorrectly, indicating that the ". . . smoke has more density" than "clean air."

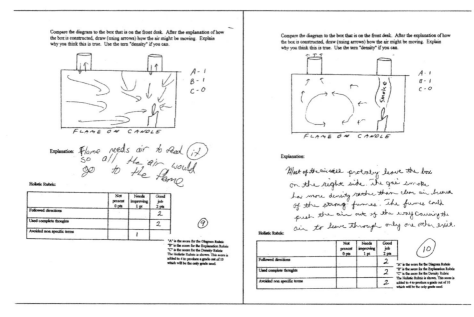

Figure 2a. Student Work Example from   Figure 2b. Student Work Example from
the Diagnostic Assessment   the Diagnostic Assessment

DEVELOPING SUMMATIVE ASSESSMENTS

I incorporate the diagnostic assessment task into my summative assessment task to provide a similar measure of student performance. Although some measurement specialists may criticize this procedure, I find that it provides me with an easy means to compare student responses. It is also useful to have other assessment tasks that will show transference of knowledge or skills to other situations (Costa, 1984). This means that the summative assessment should provide the students with more challenging situations. I also include other supportive concepts previously introduced during the unit. In my unit on convection it was important that students understand how air is heated in the atmosphere to set up convection currents, and the cause-effect relationship that results in high and low pressure areas. Therefore, the summative assessment needs to be more comprehensive, but not so complicated that scoring would become cumbersome.

As with the diagnostic assessment, the purpose of a summative assessment must be defined first. I aim to evaluate individual student performance in order to produce a fair score and grade and to evaluate the effectiveness of my teaching during the unit. To this end, I always use some selected response items as well as a variety of alternative assessment tasks to produce a manageable but effective and fair assessment tool. I include the same task used on the diagnostic assessment, and I develop a similar task that requires the transference of relevant concepts. A student response to the first task is shown in Figure 3. The second task shows an enclosed box with an ice cube that would cool the air. Student responses to this task are shown in Figure 4. I used the same rubrics for these tasks as I had used on the diagnostic task. I also included an authentic situation: a desert and forest area being heated by the sun. For this task, three prompts guided students thinking and responses. Student responses to this task are shown in Figure 5. This task required the understanding that the air is being heated by the earth's surface at different rates, resulting in low and high pressure areas. The analytic rubric applied to this task addressed these issues (Table 3).

The student responses were also assessed using the holistic rubric for response format. A summative assessment included open-ended response tasks with appropriate scoring rubrics, along with selected response items (e.g., multiple-choice, matching, and true and false questions) and skills (e.g., graphing and interpretation of graphs and data tables). These align with instructional goals and create a score that may be used in the grading process.

Table 3. The Analytic Rubric for the Summative Assessment

Diagram Rubric

Student showed no arrows	0 pts.
Student showed incorrect or confused motion	1 pt.
Student showed air rising over the desert or air sinking over the forest but no resulting motion elsewhere	2 pts.
Student showed most motion but not complete motion	3 pts.
Student showed a convection current with air rising over the desert and air sinking over the forest with air moving between the areas	4 pts.

Prompt #1 Analytic Rubric

Student has no answer or answer does not involve the sun or surface	0 pts.
Student states that the sun alone, or the surface alone, heats the air	1 pt.
Student states that the sun heats the surface and the surface heats the air	2 pts.

Prompt #2 Analytic Rubric

Student has no answer or incorrect answer	0 pts.
Student states that the low pressure area is over the desert or that the high pressure is over the forest	1 pt.
Student states that the low pressure area is over the desert and that the high pressure area is over the forest	2 pts.

Prompt #3 Analytic Rubric

Student has no answer	0 pts.
Student response does not include the idea that heated air expands and becomes less dense	1 pt.
Student response includes the idea that heated air expands and becomes less dense	2 pts.

Examples of Student Responses

The student response in Figure 3 shows that the diagram correctly illustrates the movement of air in and out of the convection box (a score of 4 on the diagram rubric). The explanation indicates that the convection current is created by the candle flame that heats the air, causing the air to rise and exit the convection box (a score of 4 on the convection rubric). The explanation, however, inaccurately describes cool air as less dense and warm air as more dense (a score of 1 on the density rubric).

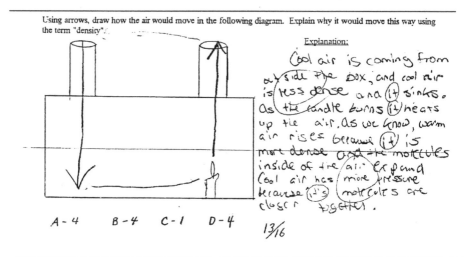

Using arrows, draw how the air would move in the following diagram. Explain why it would move this way using the term "density".

Explanation:

Cool air is coming from outside the box, and cool air is less dense and it sinks. As the candle burns it heats up the air. as we know, warm air rises because it is more dense and the molecules inside of the air expand. Cool air has more pressure because its molecules are closer together.

A - 4   B - 4   C - 1   D - 4

13/16

Figure 3. Student Work Example from the Summative Assessment

The ice cube and candle tasks are scored similarly. One difference is that the convection current created by the ice cube cools the air near the ice, increasing its density and causing it to sink. As the air moves away from the ice cube, it warms and rises. This process of heating and cooling continues until the ice cube melts and the air temperature inside the box reaches a state of equilibrium. The two student responses in Figure 4 illustrate an excellent response (Figure 4a) and a poor response (Figure 4b). The student with the excellent response drew arrows that accurately depicted the convection current (a 4 on the diagram rubric). His explanation appropriately indicated that cool air is denser than warm air (a 4 on the convection rubric and a 2 on the density rubric). The student response that reflects a poor performance drew arrows that inaccurately represented the movement of air (a 1 on the diagram rubric) and the explanation focused on the warm air melting the ice cube (a 1 on the convection rubric and a 0 on the density rubric).

Using arrows, draw how the air would move in the following diagram. This is a box with no openings. Explain why it would move this way using the term "density".

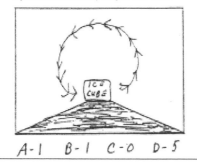

Explanation:

the ice cube is cooler than the rest of the bottom of the box so the air just above the ice cube will cool the air above and the air will become More dense so it drops. While the air above the bottom of the box will get warn and rise etc.

A-4   B-4   C-2   D-5   15/16

Figure 4a. Student Work Example from the Ice Cube Task

Using arrows, draw how the air would move in the following diagram. This is a box with no openings. Explain why it would move this way using the term "density".

Explanation:

The air is making the box more dens making warner melting the Ice cube melt

A-1   B-1   C-0   D-5   7/16

Figure 4b. Student Work Example from the Ice Cube Task

The two student responses for the forest and desert task also illustrate an excellent (Figure 5a) and a poor (Figure 5b) response. The excellent response included arrows that accurately represented air movement (a 4 on the diagram rubric). For the first prompt, the student explained that the ". . . sun warms the earth's surface which warms the air" (a score of 2). On the second prompt, the student indicated that "The forest would have high pressure and the desert would have low pressure" (a score of 2); for the third prompt, the student indicated that warm air is less dense than cool air (a score of 2). The second student's response did not contain arrows accurately illustrating the movement of air. His response to the first prompt indicated that only the sun heats the air. The reply to the second prompt was incomplete; while the response to the third prompt did not include the idea that heated air expands and becomes less dense.

On the following diagram, place arrows that will show how the air will move because of the sun's effect on this system. Show where the air will rise and sink and show any resulting surface wind.

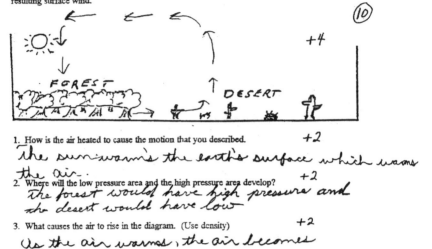

1. How is the air heated to cause the motion that you described.   +2

The sun warms the earth's surface which warms the air.

2. Where will the low pressure area and the high pressure area develop?   +2

The forest would have high pressure and the desert would have low

3. What causes the air to rise in the diagram. (Use density)   +2

As the air warms, the air becomes less dense and rises.

Figure 5a. Student Work Example from the Forest and Desert Task

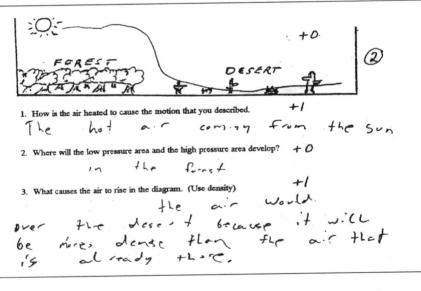

1. How is the air heated to cause the motion that you described.   +1

The hot air coming from the sun

2. Where will the low pressure area and the high pressure area develop?   +0

in the forest

3. What causes the air to rise in the diagram. (Use density)   +1

the air would. over the desert because it will be more dense than the air that is already there.

Figure 5b. Student Work Example from the Forest and Desert Task

CONCLUDING THOUGHTS

Sometimes alternative assessments are used only to assess process and performance skills. These are important, but alternative assessment formats can also be successfully used to identify students' conceptual understandings and changes in their understandings. Diagnostic and summative assessment techniques are needed to fully assess student progress and instructional effectiveness.

By experimenting with alternative forms of diagnostic and summative assessments, I have gained new insights about the conceptual levels of my students. I have gained a better sense of the effectiveness of my instruction, and this has made a positive impact on my teaching. Perhaps the most impressive result is that I have seen my students learn to better express their ideas more clearly and concisely than I had ever thought possible.

REFERENCES

American Association for the Advancement of Science. (1990). *Science for all Americans*. New York: Oxford University Press.

American Association for the Advancement of Science. (1993). *Benchmarks for Science Literacy*. New York: Oxford University Press.

Costa, A.L. (1984). Thinking: How do we know students are getting better at it? *Roper Review*, 6(4), 197-199.

Cizek, G.J. (1997). Learning, achievement, and assessment: Constructs at a crossroads. In G.D. Phye (Ed.), *Handbook of Classroom Assessment: Learning, Adjustment, and Achievement* (pp. 2-32) San Diego, CA: Academic Press.

Good, T.L. & Brophy, G.E. (1986). *Educational Psychology* (3rd edition). New York: Longman.

Wittrock, M. (1977). *Learning and Instruction*. Berkeley, CA: McCutchan.

Wolf, A., Bixby, J., Glenn, J., & Gardner, H. (1991). To use their minds well: Investigating new forms of student assessment. In G. Grant (Ed.), *Review of research in education* (pp. 31-74). Washington, DC: American Educational Research Association.

13. ALTERNATIVES TO TEACHING AND ASSESSING IN A HIGH SCHOOL CHEMISTRY CLASSROOM: COMPUTER ANIMATIONS AND OTHER FORMS OF VISUALIZATION

Why use alternative assessment? Why doesn't traditional assessment assess students' understandings and knowledge just as well? Perhaps my experience at Christmas one year best answers these questions. While shopping at a local mall I heard from across the store; "Mrs. Flick, Mrs. Flick, How are you?"

In response to my question about how the first year of college was going, Heather answered enthusiastically "Great! Chemistry was so easy! When we got to the parts of chemistry that we didn't cover in high school all my friends asked, 'How can you do so well?' I answered, 'Don't you see those pictures in your mind? Didn't your chemistry teacher teach you how to make pictures in your mind?'"

Science is a process, not just "a bag of facts" (Hurd, 1991, p. 34). As Hurd puts it: "Traditional discipline-bound, fact-laden science courses are too narrow in scope to do this . . . The goal is not to predict the future but to use what we have learned to plan and direct the future" (1991, p. 34).

In fact, meaning is created in the mind of the student as a result of sensory interactions with the world (Saunders, 1992). This means that the teacher cannot simply transfer knowledge to the student's mind. Understanding may be constructed through and reflected in images (Gagne & White, 1978). Images are the mental representation of sensory perceptions that are often visual or otherwise related to the senses (White & Gunstone, 1992). Images represent the learner's visualization of the concepts in their mind, their constructed meanings of the concepts. It has been my experience that using alternative forms of assessment not only leads to a better evaluation of students' understandings and knowledge, but also to alternative teaching methods that lead to the more meaningful learning of science. Further, my teaching is conceptually and inquiry driven, therefore traditional multiple-choice tests do not adequately or fairly assess my students' understandings and knowledge.

In this chapter I describe two chemistry units I developed using alternative teaching and assessment methods as well as my revision of these units based on assessment results. The units emphasize computer animation and other forms of visualization as teaching and assessing techniques. Both units are based on the *Benchmarks for Science Literacy* (American Association for the Advancement of Science [AAAS], 1993). The Physical Setting, Structure of Matter (p. 80) states that by the end of the twelfth grade, students should know that:

211

Daniel P. Shepardson (ed.), Assessment in Science, 211—225.
© 2001 *Kluwer Academic Publishers. Printed in the Netherlands.*

1. Atoms are made of a positive nucleus surrounded by negative electrons. An atom's electron configuration, particularly the outermost electrons, determines how the atom can interact with other atoms. Atoms form bonds to other atoms by transferring or sharing electrons.

2. Atoms often join with one another in various combinations in distinct molecules or in repeating three-dimensional crystal patterns. An enormous variety of biological, chemical, and physical phenomena can be explained by changes in the arrangement and motion of atoms and molecules.

3. The configuration of atoms in a molecule determines the molecule's properties. Shapes are particularly important in how large molecules interact with others.

The first unit about chemical reactions is based on a series of laboratory investigations to introduce the concepts of chemical shorthand for chemical reactions (chemical symbols, formulas, and equations) and on classification of chemical reactions as to type in regard to kind and energy. These concepts are reinforced through the use of teacher-made computer animations based on the laboratory investigations conducted by the students. It involves alternative assessment in which students plan and conduct their own laboratory investigations, write a laboratory report, and generate a computer animation about one of the chemical reactions that occurred in the laboratory investigations. The second unit emphasizes chemical bonding at the atomic level and involves computer animations to teach molecular shapes. For the alternative assessments, students construct concept maps, build molecular shapes using marshmallows and toothpicks, and generate computer animations that illustrate bonding at the atomic level.

The overarching instructional goal of my teaching is to make the abstract atomic level of chemistry more concrete and visual to students, to enhance their conceptual understanding and knowledge construction. My classroom assessment practice is designed to assess students' conceptual understanding and to enhance the learning process, by creating assessment activities that are also learning experiences. The specific goals of my assessment techniques are as follows:

1. Concept maps: to assess students' conceptual understanding of the connections between the concepts involved in the unit and to assist students in seeing the subtle connections between concepts.

2. Resolution laboratories: to assess students' science process and higher level thinking skills by having them plan and conduct laboratory investigations that require the application of chemistry concepts from the series of laboratory experiences in order to interpret and explain results.

3. Written laboratory report: to assess students' ability to communicate their interpretations of the laboratory investigations using the appropriate chemical shorthand and classification scheme.

4. Student generated computer animation: to assess students' visualization and representation of the atomic level interactions that occur during the chemical reactions in the resolution laboratory investigations. This indicates understanding of the use of chemical shorthand and classification.

5. <u>Construction of molecular shapes from marshmallows and toothpicks</u>: to assess students' ability to visualize molecular shape and chemical concepts in three dimensions.

UNIT ONE: CHEMICAL REACTIONS, CLASSIFICATION, AND REPRESENTATION

I began to change my teaching and assessment approach in order to make constructivism a referent in the students' learning (Tobin & Tippins, 1993). For me, constructivism as a teaching referent requires that students first be provided with instructional activities that involve making observations, collecting data, and constructing interpretations based on prior conceptual understanding and knowledge. Students then discuss these observations, data, and interpretations, as well as the scientific concepts for explaining the phenomenon in a different way. Next, students might read about the concepts and conduct different laboratory investigations using their new understanding and knowledge to explain their results.

My new approach to teaching about chemical reactions began with experiences in observing and manipulating equipment, materials, and phenomena. This provided a basis for students to make their understandings explicit. I had to devise a series of experiences that would demonstrate the chemistry concepts that I wanted to teach. Because students construct their own understandings and use their prior understandings to see, interpret, and explain events, I realized that students might not view the series of experiences in the same way I would or a chemist would, but I had to start somewhere.

I started by constructing an outline of the chemistry concepts that I wanted students to understand. I reviewed the *Benchmarks*, *Science for All Americans* [SFA] (AAAS, 1990) and my school's curriculum guide. In the *Benchmarks* and SFA I drew directly from the sections on "Scientific Inquiry" and "The Physical Setting." I found that my school's curriculum guide specified a list of concepts to cover. A concept map helped me to identify the sequence of the laboratory experiences. The concept map forced me to think about what I wanted students to learn. The concept map also helped me to see connections between concepts that I otherwise might have overlooked. Based on the concept map, I selected and sequenced the following laboratory investigations:

1. Heating magnesium
2. Calcium oxide plus water
3. Heating copper (II) carbonate
4. The reaction of magnesium and hydrochloric acid
5. The reaction of various metals with hydrochloric acid
6. The reaction of copper (II) sulfate solution and sodium hydroxide solution
7. The reaction of sodium hydroxide solution and hydrochloric acid

The students were given minimal instruction about the laboratory investigations and were directed to record what they considered pertinent data. Following the first laboratory, the class discussion emphasized the observations and the data collected, as well as students' interpretations. The students then read the appropriate chapters from the textbook. The next day the students viewed my computer animation of what occurred on the atomic level during the first laboratory investigation and the chemical language and classification used to represent the chemical reaction. While viewing the animation, the students completed an accompanying worksheet. The animation was made using *Macromedia Director*, which is compatible with Macintosh computers. This order of activity makes the abstract representation of chemical reactions more concrete and visual. The students then performed the remaining series of laboratory investigations and viewed my animation about the heating copper (II) carbonate laboratory investigation.

The unit concluded with a summative assessment that required students to select one of the remaining laboratory investigations and construct a computer animation to demonstrate their understanding and knowledge of chemical reactions, chemical formulas, and chemical equations. This assessment task encouraged students to construct knowledge and represent it in a visual manner using computer animation. It required students to talk to each other about the laboratory investigations, their understandings of the laboratory investigations, and how to use the computer animation program as a tool for representing their understandings. This assessed students' conceptualization of what occurs during a chemical reaction and how the chemical reaction is represented thorough formulas and equations--chemistry's language. The interactions between students required a decision-making process in order to make the finished product. The student-student discussions often led to argumentation, requiring students to defend their ideas and understandings. The assessment task not only assessed students' understanding and knowledge about chemical reactions, but also students' interpersonal skills and team work. Science concepts were represented visually, technology was being used, and cooperation was being indirectly taught.

The Macromedia animation program involved three parts: the paint window, the score, and the stage. The first day, students were taught to use the paint window to represent compounds with accompanying formulas, as experienced in Laboratory Investigation One. These molecular representations were then assessed using the scoring guide provided to the students at the time they programmed their molecular animations; this was a part of the instructional process. The next day students were taught to use the score (i.e., the window where the animation is planned) and the stage (i.e., the window where the animation takes place), using the molecular animations they had programmed. The animation represented the chemical equations from Investigation One, and was assessed with the scoring guide (Table 1). The students were allowed to select the reaction they wished to animate.

The process used to teach the students how to use the animation program required them to practice visualizing and representing molecules, how the molecules interact in chemical reactions, and to use the chemical shorthand to represent molecules and chemical reactions. This makes the use of time to teach the animation process valuable to the teaching of chemistry. The students' animations for the summative assessment demonstrated whether they understood chemical reactions. For example, a student who could balance an equation using coefficients might not have understood the concept of chemical reactions beyond the mathematics involved in balancing equations. Such a student would have no concept of atoms and molecules moving and hitting with enough energy to cause a chemical reaction.

Table 1. Chemical Reaction Computer Animation Assessment Task and Scoring Guide

Assessment Task
Using the substances created yesterday, construct simple animations to demonstrate the three chemical reactions listed below. Title each reaction with the appropriate number. Each animation should be at least 20 frames long. The animations must occur one after another in the file. Remember to use the three commands demonstrated. Put the word "equation" and COMPLETE balanced chemical equation on the final frame of each animation; also indicate the type of chemical reaction and the type of energy change.

Reaction 1. Solid sodium plus zinc (II) chloride solution. The temperature of the solution rises.
Reaction 2. Beryllium chloride solution plus calcium phosphate solution. Temperatures of the beginning solutions are 24 C. Temperature of the final mixture is 22 C.
Reaction 3. Calcium carbonate solid plus heat.

Scoring Guide
Each reaction will be scored based on the guide below:
_____ States of matter (8 points)
_____ Number of reactants (4 points)
_____ Number of products (4 points)
_____ Word equation (2 points)
_____ Balanced chemical equation (4 points)
_____ Classification of reaction (2 points)
_____ Energy change (2 points)

The following student animation on sodium hydroxide plus hydrochloric acid is well done. The student errors were related to forgetting details associated with balanced equations. These errors were not associated with the visualization process. On the other hand, one group (Sarah and Tricia) did not make the connection between the balanced equation and the number of reactants and products on the screen. Sarah and Tricia did not connect the visualization with the chemical shorthand, which is a serious error and could indicate that the visualization process did not add to their understanding as I had hoped. Two other animations show the same type of errors. One group received a low score for not following directions;

these students did not add formulas or names to the animation. On the whole, however, the students were very successful at visualizing and representing what occurred on the atomic level during the chemical reaction done in the laboratory.

I have encountered two drawbacks to this assessment: time and money. Students must learn to use computers and the animation program. On the other hand, computers will be a part of are future. Even though students may not use this particular animation program in the future, many of the techniques involved are transferable to other computer programs. In terms of cost, a teacher who does not have access to computers or software could use a flipbook such as those used in art class to make cartoons. The only materials required would be paper and colored pencils. This form of assessment with flipbooks would not show the same motion, but it would still require higher level thinking and application of understanding and knowledge of chemical reactions.

Revising the Unit

The following year I modified the process by first conducting Laboratory Investigation One followed by student viewing and discussion of the animations generated by students from the previous year. Students also viewed my animations of Laboratory Investigations Three, Four, and Six. Students then wrote short laboratory reports about all seven investigations. The students and I then generated a concept map that organized and structured the concepts involved in the investigations and covered in the readings. To do this, students first constructed their own concept maps and then worked in small groups of four to make a group concept map that was assessed with a scoring guide. I collected these and then directed a whole-class construction of a concept map, sharing my ideas about the various connections and interrelationships.

Students then set up the "One Tube Reaction" summative assessment task. This involved everyday chemicals and produces a series of visible chemical reactions (see Appendix A for the directions for the "One Tube Reaction"). The students gathered data about the reactions for five days. At the end of this time a class discussion elaborated the conclusion section of the formal laboratory report. I allowed student questions to direct the discussion and I became the coach, not the lecturer. This discussion took approximately two days. Students then made a computer animation of one of the four reactions that occurred in the "One Tube Reaction" task.

The second year's assessments showed that the students seemed to have a better understanding of the relationships between chemical language and atomic level reactions. The coefficients and the number of molecules were most often in line. The animations revealed more subtle errors such as those shown in the iron dissociating animation (Figure 1). Interestingly, the student teams did well on the animations but poorly on the written report in terms of demonstrating the connection between chemical language and the atomic level.

This may be a result of the fact that the animations were done after the written laboratory reports because of the availability of the computer laboratory. Next time, I want to reverse the process to see if it will affect the quality of the written laboratory report. As I work with alternative assessment I continue to revise and refine both my teaching and assessment. Next year, I will use the revised scoring guide displayed in Table 2 to more easily assess the data chart section of the written report. This may also improve communication with students when the written report is returned.

I have found scoring guides help in the following ways:

1. They help to organize my thoughts (e.g., the concept maps). What do I want the students to know?
2. If handed out prior to the assessment, scoring guides enhance communication by increasing the number of questions asked by students. They also make the students more comfortable with the assessment task. Students seem relieved because they know what is expected.
3. When handed back after assessment, scoring guides improve communication because students can identify where they need improvement.
4. Scoring guides make it easier and faster for me to assess students' work because the criteria are constantly before me. I am able to be more consistent in assessing students.

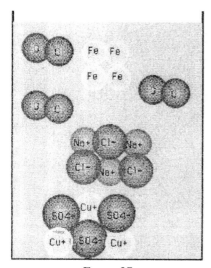
Frame 27

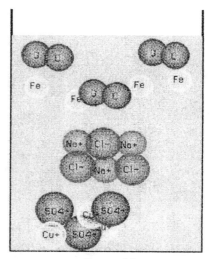
Frame 36

Figure 1. Sample Student Computer Animation (Frames) Illustrating Subtle Errors in the Iron Dissociating in the Chemical Reaction

Table 2. Revised Scoring Guide/Form for Laboratory Report

Part I. Title, Purpose, Procedure, Data Chart One

Category	Score
Title (2 pts.)	
Purpose (2 pts.)	
Procedure (2-8 pts. Sig. Dig., 2 pts. sentences, 4 pts. amounts)	
Data chart of observations (20 pts.)	
First day (Drawing + Description 4 pts.)	
Second day (4 pts.)	
Third day (4 pts.)	
Fourth day (4 pts.)	
Fifth day (4 pts.)	

Part II. Data Chart Two, Observations of 14 Vials

Name	Formula (2pts each)	Appearance (2pts each)	Present (2pts each)
Copper metal	Cu		
Copper (II) chloride solid	$CuCl_2$		
Solid copper (II) hydroxide as a precipitate	$Cu(OH)_2$		
Solid copper (II) oxide	CuO		
Iron (III) hydroxide solution	$Fe(OH)_3$		
Solid iron (III) chloride	$FeCl_3$		
Solid iron (III) oxide	Fe_2O_3		
Solid iron (II) sulfate	Fe_2O_3		
Hydrochloric acid	HCl		
Solid mercury (II) iodide	HgI_2		
Solid potassium chromate	K_2CrO_4		
Sodium hydroxide solution	NaOH		
Sodium sulfate solution	Na_2SO_4		
Solid zinc	Zn		

Part III. Conclusions, discussion of five days of observations (Describing change, what you think the chemical is, and what is formed.

Day	Description (2pts each)	Chemical (?) (2pts each)	Form (?) (2pts each)
1			
2			
3			
4			
5			

Table 2 Continues

Table 2. Revised Scoring Guide/Form for Laboratory Report, Continued

Part IV. Detailed Analysis of Two Reactions

Item	Reaction One	Reaction Two
1. Word equation (4 pts.)		
(a) Words (2 pts.)		
(b) States (1 pt.)		
(c) Energy (1 pt.)		
2. Circles (24 pts.)		
No. circles in reaction (2 pts.)		
(a) Size of circles (1 pt.)		
No. of circles in molecule (1 pt.)		
(a) Labeled (2 pts.)		
1. Symbol (1 pt.)		
2. Oxidation no. (1 pt.)		
3. Chemical equation (12 pts.)		
(a) Coefficient (3 pts.)		
(b) Formula (3 pts.)		
(c) State (3 pts.)		
(d) Energy (3 pts.)		
4. Classification (5 pts.)		

Part V. Applications

Item	Reaction One	Reaction Two
1. Word equation (4 pts.)		
(a) Words (2 pts.)		
(b) States (2 pts.)		
2. Occurrence in world (6 pts.)		
(a) Where (2 pts.)		
(b) How (4 pts.)		
3. Effect on the world (4 pts.)		

One problem with assessing concept maps, however, was that the scoring guide included too many points and a new scoring guide had to be created each time. I plan to design a more generalized scoring guide and will field test the following format next time:

0-4 points: all concept words used
0-4 points: proper number of hierarchy levels
0-4 points: proper branching of concepts
0-4 points: connecting words used on all levels
0-4 points: proper number of interconnections shown
0-8 points: connections are logical

I will also add a section to the conclusion of the written report that requires students to select two of the chemical reactions in the test tube assessment task and discuss examples of these reactions in everyday circumstances.

UNIT TWO: CHEMICAL BONDING, VALENCE ELECTRONS, TYPES OF BONDS, AND MOLECULAR SHAPES

In the second unit, I used computer animations to teach and assess students' conceptual understanding. I designed two summative assessments: one was a student-generated animation on the atomic level. This assessed the students' understanding of the bonding process, valence electrons, and the types of bonds. The second followed the computer animation about molecular shapes in which students were asked to construct models of molecules using marshmallows and toothpicks. This assessment task assessed students' ability to start with a formula and produce a three-dimensional model of the substances. This involved many levels of knowledge: writing formulas, drawing Lewis Dot representations, predicting the appropriate molecular shape, and creating three-dimensional models of molecules.

Working in pairs, students made animations about the bonding of three different combinations of elements to demonstrate the formation of the three types of chemical bonds: ionic, polar covalent, and non-polar covalent. Again, the student-generated animations revealed common misconceptions. While making the animations, over 50% of the class used all of the beryllium electrons to form the bond with carbon, resulting with one beryllium combined with one carbon, rather than two beryllium. This demonstrated that even though the students had read, heard a lecture, and completed homework exercises concerning valence electrons, they did not understand that the inner two electrons of Beryllium would not take part in the bonding process (Figure 2).

The other misconception made visible by this alternative assessment was the difference between the three types of bonds. The students missed the point that covalent bonding involves the sharing of electrons, while ionic bonding involves a complete transfer of electrons. The difference in the types of covalent bonding are a matter of the time the electrons spend around the individual nuclei. Students did not realize that this was demonstrated by the number of frames an electron revolved around particular nuclei in the animation.

This assessment revealed to me, that traditional methods of teaching (e.g., read, lecture, and do homework) had not really taught the students. The students did not have the proper pictures in their minds; they did not understand. I went back to the drawing board and revised my teaching, but not my assessment. I did, however, revise the scoring guide (Table 3).

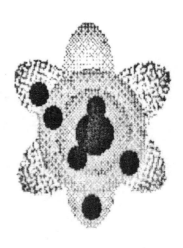

Frame 53

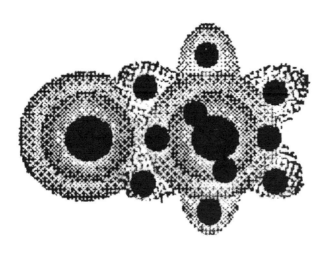

Frame 54

Figure 2. Sample Student Computer Animation (Frames) Illustrating the Misconception that the Inner Two Electrons of Beryllium Take Part in the Bonding Process

Table 3. Revised Scoring Guide for Assessing Students Animations

Reactants	Product
Structure (2pts.)	Type of bond (4 pts.)
Number of electrons (2 pts.)	State of matter (2 pts.)
Movement of electrons (4 pts.)	Movement of core electrons (2 pts.)
Correct number of reactants (4 pts.)	Movement of valence electrons (2 pts.)
	Number of ions (4 pts.)
Process of Bonding	
Movement of reactants (2 pts.)	**Labeling on last frame**
Movement of electrons (2 pts.)	Word equation (4 pts.)
Bond length established (6 pts.)	Lewis Dot Representation (8 pts.)
(a) Step one (2 pts.)	(a) Reactants (4 pts.)
(b) Step two (2 pts.)	(b) Product (4 pts.)
(c) Step three (2 pts.)	Type of bond formed (2 pts.)
	Name of compound (2 pts.)
	Formula for the compound (4 pts.)

    The next year I approached this topic from a constructivist perspective. Granted it is difficult for students to actually do a laboratory investigation in which they see electrons move during chemical bonding, but I could approach the process differently. This time I provided students with a list of terms to define. I told them about the textbook chapter that we would be studying and left it up to them to decide how to accomplish the task of defining the terms. The next day, I asked if there were any questions about the definitions. There were very few. I then showed students a computer animation about the three types of chemical bonds. The students completed a worksheet about the animation and used their list of terms to assist in completing the worksheet. Students worked in pairs, and I observed much discussion between partners. After everyone had finished the worksheet, we had a class discussion about the answers for the worksheet, and we reviewed the computer animations. Based on the student-generated computer animations, there were few students who made the mistake of combining beryllium and carbon.

    The second summative assessment of this unit required the students to construct models of molecules using marshmallows and toothpicks. This assessment task occurred after students had completed textbook reading, homework exercises, and after they had viewed computer animations demonstrating molecular shapes in three dimensions. This assessment demonstrated the incomplete conceptual understanding students held about three-dimensional shapes. Despite the computer animations and my own demonstration using balloons to emphasize the three-dimensional shape of molecules, many of the students still represented the structure of the tetrahedral, trigonal pyramidal, and trigonal bipyramidal in two dimensions as in the textbook. Perhaps we should throw the textbook away--at least the two dimensional chart. I again went back to the drawing board.

Revising the Unit

The next year, I tried an alternative teaching method. This time, students read the section in the textbook involving molecular shapes and then constructed concept maps. Students viewed the computer animations on molecular shapes, and produced a group concept map incorporating information from the animations. These teams then presented and explained their concept maps to the class. A class concept map was constructed and posted on the wall. Students then built models of the molecular shapes with marshmallows and toothpicks. The directions for this year were the same, however, I changed the scoring guide (Table 4). Some students still had difficulty with the tetrahedral shape and still produced the two-dimensional shape; however, alternative teaching led to greater comprehension.

Table 4. Scoring Guide for the Marshmallow Molecular Models Task

Formula (2 pts.)
Name of the particle (2 pts.)
Electron dot diagram (4 pts.)
Number of atoms bonded to the central atom (2 pts.)
Lone pairs of electrons (2 pts.)
Predicted molecular shape (2 pts.)
Size of marshmallows used for ions (4 pts.)
Actual shape of the molecule (4 pts.)
Makes a molecule (3 pts.)

Student Evaluation of Unit Two Teaching and Assessing

After Unit Two, my students filled out an evaluation sheet about the unit. The results indicated that students preferred instructional activities that caused them to *do* things instead of just reading. Only one student ranked reading highly. This student indicated that he liked to read. Many students commented that the reading was too confusing; they did not understand the words. The teacher made animations received the highest marks and most comments. The student-made animations also generated comments such as, "seeing and making caused me to understand," "Drilled idea into my head," "Made me think more," "Don't like the animation program," and "Didn't have enough time."

Students liked the computer animation assessment tasks the most and building molecular models the least. Comments about the multiple-choice sections were, "Can guess," "Allows guessing," "Choices help remember," and "May know the answer, but don't know how to word the answer." The multiple-choice questions seemed to allow students to guess the answers. The comments about the alternative assessment tasks were, "Hate essays," "Truly tests understanding," and "Lets me draw pictures in my mind."

CONCLUDING THOUGHTS ABOUT ALTERNATIVE ASSESSMENT IN SCIENCE

In my experience, alternative teaching and assessing methods that involve visualization and representation are worthwhile and meaningful activities for students. Student assessment results are better and information about students' conceptual understanding and knowledge are made more visible than they are through traditional assessment approaches. Alternative methods of teaching and assessing exposed misconceptions that perhaps would not have been uncovered in more traditional classrooms. Students also approved of the alternative methods of teaching and assessing. This has strengthened my belief in the need to teach with methods that encourage visualization and representation of concepts (e.g., concept maps, laboratory activities, computer animations, or model construction). My experience with Heather, as described at the beginning of the chapter, supports my view that it is not the quantity of information covered that is important, but how it is covered that matters.

REFERENCES

American Association for the Advancement of Science (1990). *Science for All Americans.* New York: Oxford University Press.

American Association for the Advancement of Science (1993). *Benchmarks for Science Literacy.* New York: Oxford University Press.

Gagne, R.M. & White, R.T. (1978). Memory structures and learning outcomes. *Review of Educational Research*, 48, 181-222.

Hurd, P.D. (1991). Why we must transform science education. *Integrating the Curriculum*, 33-35.

Saunders, W.L. (1992). The constructivist perspective: Implications and teaching strategies for science. *School Science and Mathematics*, 92(3), 136-40.

Tobin, K. & Tippins, D. (1993). Constructivism as a referent for teaching and learning. In K. Tobin (Ed.), *The practice of constructivism in science education* (pp. 3-22). Washington, DC: AAAS Press.

White, R. & Gunstone, R. (1992). *Probing understanding.* London, England: Falmer Press.

APPENDIX A: DIRECTIONS FOR THE ONE TUBE REACTION

Purpose

The goal of this laboratory is to allow you to observe changes that take place in the one test tube system in order to practice your observation skills, data-keeping techniques, ability to predict what changes are occurring in the tube, practice writing word equations, draw circles to represent the reaction, practice writing complete balanced equations, and practice classifying reactions.

Materials

One large test tube, one piece of filter paper, one iron nail with the finish removed, two grams of copper (II) sulfate crystals, four grams of sodium chloride, sheet of parafilm, and a pair of scissors.

Procedure

WEAR APRON AND GOGGLES
1. Place the copper (II) sulfate crystals in the bottom of the test tube.
2. Cut out a filter paper circle to fit the inner tube's diameter and place it on the top of the crystals.
3. Slowly, with minimal disturbance, add enough water to cover the copper (II) sulfate crystals and the circle.
4. Carefully pour the sodium chloride on top of the filter paper circle.
5. Place a second filter paper circle over the sodium chloride.
6. Carefully add water to cover the slat and then add water so that there will be about 2 inches of water above the filter paper.
7. Place the iron nail in the water on top of the filter paper circle. Make sure you have sanded it to remove the finish. Make sure it is completely covered by the water.
8. Cover the test tube with the parafilm.
9. Label your test tube and place it in your drawer.
10. Record all the changes that take place for the next five days using pictures and words.
11. Observe the 14 vials of chemicals that are placed around the room as possible suggestions of what products may have formed in the test tube. Make a data table about these vials with the following headings—name, formula, description of the chemical in the vial, and prediction if this chemical could be present in the test tube. List the chemicals in the table in alphabetical order.

DAVID E. EMERY

14. AUTHENTIC ASSESSMENT IN HIGH SCHOOL SCIENCE: A CLASSROOM PERSPECTIVE

Authentic assessment provides students with situations that engage their science process and inquiry abilities and employ their science understandings in the context of solving problems. The development of a scoring guide helps the teacher to clearly define the performance categories or science domains to be used in assessing students' performance on the authentic assessment task. Eventually, I came to realize that students, when provided with the scoring guide, would seek to improve their performance and product simply in order to meet the criteria of the scoring system. For the students, this resulted not only in improved levels of understanding, but also in increased acceptance of ownership for their performance, product, and grade. In this chapter, I illustrate my use of authentic assessment tasks, scoring guides to guide students' activity, self and peer assessment, and my assessment of students' performance. I will also share elements of my own personal growth in using authentic assessment tasks and my ideas for developing such tasks. I first provide a brief overview of what counts as authentic assessment.

WHAT IS AUTHENTIC ASSESSMENT?

Recently, authentic assessment has received much attention in the science education and assessment literature. It has been presented as the *crème de la crème* of classroom assessment, but what constitutes authentic assessment in science classrooms? In this section I present a brief overview of the literature to better define authentic assessment in science. The overview also provides background information for reflection on the strengths and weaknesses of my own assessment practice. Some readers may not agree that my assessment tasks are authentic, but they reflect a teacher's perspective on what authentic assessment looks like in a science classroom.

Based on the educational literature, a key aspect of authentic assessment is that it involves students in completing tasks that are worthwhile, significant, and meaningful (Newmann, 1991). Authentic assessment occurs in a meaningful context when it relates to authentic concerns and problems encountered by students, requiring students to exhibit what they have learned through the application of knowledge and skills (Brooks & Brooks, 1993).

Daniel P. Shepardson (ed.), Assessment in Science, 227—247.

The *National Science Education Standards* (National Research Council [NRC], 1996) describe authentic assessment as:

> . . . exercises that closely approximate the intended outcomes of science education. Authentic assessment exercises require students to apply scientific knowledge and reasoning to situations similar to those they will encounter in the world outside the classroom, as well as to situations that approximate how scientists do their work (p. 78).

The NRC (1996) standards continue by providing an example of an authentic assessment task on toxic waste:

> An alternative and more authentic method is to ask the student to locate such information and develop an annotated bibliography and a judgment about the scientific quality of the information (p. 84).

The "Teaching Standards" section of the NRC (1996) standards provides an example of an assessment activity titled the "Science Olympiad" that is identified as an authentic assessment for grades one through four. The assessment activity requires students to complete different tasks at four different stations; for example, at Station B, Rolling Cylinders, Task 1 requires students to:

1. Roll each cylinder down the incline.
2. Describe the motion of the cylinders and their relation to each other (NRC, 1996, pp. 39-40).

The question one might ask is "Do these assessment tasks reflect what scientists do or what students would do outside of school in the real world?" Or, "Do these tasks reflect authentic assessment in terms of what students do in science classrooms?"

Professionals in the field, including scientists, produce rather than reproduce knowledge, express knowledge through discourse, create products based on and that reflect knowledge, and engage in performances that communicate knowledge to others (Newmann, 1991). Although children cannot perform at the same level as adults, their performance can develop in a similar direction (Newmann, 1991). Thus, Newmann (1991) contends that authentic assessment should reflect what practicing professionals do. Newmann (1991) argues that authentic assessment should reflect disciplined inquiry, featuring the use of prior knowledge, in-depth understanding, and the production of knowledge in an integrated form. It should also have aesthetic, utilitarian, or personal value apart from the value in assessing the learner. Authentic assessment requires students to engage in disciplined-based inquiry, to produce or apply knowledge, to value the experience beyond the classroom, and to produce a product or engage in a means of communicating their understandings.

Baron (1991) has identified the characteristics of "enriched" performance assessment tasks. Such tasks are:

- Grounded in real world contexts.
- Involved sustained work extended over several days combining in-class and out-of-class work.

- Emphasized the "big ideas" and major concepts in science.
- Blended essential content with essential processes, often requiring the use of scientific methodology and the manipulation of scientific tools.
- Presented open-ended and loosely structured problems that require students to define the problem and develop a strategy for solving the problem.
- Required students to determine what data are needed, to collect, report, and portray the data, and to analyze the data.
- Accompanied by explicitly stated scoring criteria related to science content, process, group skills, communication skills, and habits of mind.
- Required students to use a variety of skills for acquiring information.

To summarize, enriched performance assessments are tasks that require students to solve problems, to produce rather than recognize solutions, and to construct responses rather than simply recall facts and concepts. These skills parallel those needed for scientific inquiry: a reliance on evidence, the use of hypotheses and theories, appropriate habits of mind, the use of existing ideas and findings, and the use of quantitative or qualitative methods in the investigation of phenomena (American Association for the Advancement of Science [AAAS], 1990). Further:

> Scientific inquiry is not easily described apart from the context of particular investigations. There simply is not a fixed set of steps that scientists always follow, no one path leads them unerringly to scientific knowledge. There are, however, certain features of science that give it a distinctive character as a mode of inquiry. Although those features are especially characteristic of the work of professional scientists, everyone can exercise them in thinking scientifically about many matters of interest in every day life (AAAS, 1990, p. 4).

The characteristics of scientific work described in *Science for All Americans* (AAAS, 1990) include: the demand for evidence, the use of logic and imagination, the use of explanation and prediction, the avoidance of bias, and the non-authoritarian stance of science as a discipline. It seems then, that authentic assessment tasks in science should reflect these characteristics of scientific inquiry.

For me, authentic assessment tasks are similar to student projects; they extend the curriculum and the instructional process. The emphasis is on student learning as well as assessing student performance. It is the context of the task that makes it authentic by requiring students to solve a problem using science content, processes, and inquiry skills, requiring a final product. I design authentic assessment tasks that reflect a variety of contexts and that engage and motivate students with different interests in accordance with the National Science Education Standards, Assessment Standard D (NRC, 1996):

> Assessment tasks must be set in a variety of contexts, be engaging to students with different interests and experiences, and must not assume the perspective or experience of a particular gender, racial, or ethnic group (NRC, 1996, p. 85).

A PERSONAL PERSPECTIVE ON PROFESSIONAL GROWTH IN DEVELOPING AND USING AUTHENTIC ASSESSMENT TASKS AND SCORING GUIDES

At some point in the late seventies I decided that my students in an earth science program needed experience with a long-term project. Because this was more open-ended, the students suggested that I provide a list of expectations for each project or authentic assessment task. The first of these projects was a thirty-day moon watch. The moon phases were to be displayed as an integral part of a model incorporating a creative theme. It was necessary to indicate to the students not only the need to accurately list the phases of the moon, but also how the project was to be assessed, based on their creativity and effort. My first scoring system for the moon project was:

Title (creative and well done)	10 pts.
Correct drawing of daily moon	30 pts. (1 pt each)
Correct name of daily moon	30 pts. (1 pt each)
Design shows creativity	15 pts.
Work indicated by project	<u>15 pts.</u>
Total	100 pts

Eventually this project grew into several others. As the number of authentic assessment tasks grew, I needed to improve the scoring guide. Even with a list of the different aspects of the task upon which their grade was to be based, students still asked which sections were to be emphasized. I partitioned each section into smaller units, complete with a description and point values. The moon phase scoring guide evolved into a more detailed description of expectations for the assessment task:

<u>Title</u>

Shows creativity in design	5 pts.
Title construction shows work	5 pts.
Correct drawings of daily moon	30 pts. (1 pt. each)
Correct names of daily moon	30 pts. (1 pt. each)

<u>Design</u>

Shows creativity in design	5 pts.
Incorporate moons into design	5 pts.
Unique (compared to samples shown)	5 pts.

<u>Work</u>

Effort and care shown in constructing moons	5 pts.
Effort and care shown in creating display	5 pts.
Project shows effort beyond expectations	<u>5 pts.</u>
Total	100 pts.

This scoring system had been expanded to help students better understand what was expected of them, and what I meant by "work" or "design." I found it interesting that, with the further refinement of the work section of the above scoring system, many more students began to strive for the extra five points for achievement.

Early on, I had hoped to display a single scoring guide for all projects, making one strategy applicable to all student projects. In this system I tried to use quantifiers such as "all," "most," "some," and "little" or "none correct." Decreasing point values were assigned to each category. The resulting scoring guide took on a variety of appearances depending upon the assessment; for example, the rowboat assessment task was described as follows:

> Given a description of four people sitting in a row boat tied to a dock in a lake, students will describe what motions are involved as the people leave the boat for the dock. They need to consider the boat, the dock and the lake. Following this, they will design an experiment to test their hypothesis of motion. Their experiment will include a hypothesis (description of motions) and a procedure for the experiment.

SCORING RUBRIC:

	All	Most	Some	Little
Hypothesis has correct movement up as people depart	10	8	6	4
Procedure contains reasonable instructions	10	8	6	4
Student interview explains procedure	10	8	6	4

Results with this scoring system never reached my expectations; students did not relate to these generalized criteria. As I spent more time trying to improve the definition of these general performance levels, the students experienced more difficulty in relating to these levels. I eventually abandoned the search for the perfect generalized scoring system. It seemed that each assessment task needed its own, individualized scoring guide in order to be useful to the students. Interestingly, at a science assessment workshop not long ago, I found that teachers wanted to start with these same generalized categories in developing their scoring systems. I shared my historical perspective from my classes.

ASSESSMENT TASKS AND SCORING GUIDES

As authentic assessment began to play a major role in my classroom, I abandoned the concept of a generalized or universal scoring system. I developed analytic scoring rubrics specific to the assessment task, also made available to students at the start of the task. I found these could be grouped by type, resulting in a scoring rubric template that guided my development of the scoring system. For example, authentic assessment tasks in which students design experiments or investigations consist of similar performance categories, but contain specific criteria determined by the context of the assessment task. I found that the performance categories tended to remain stable, but the criteria changed to reflect the context of the task. In authentic assessment tasks that require students to design a laboratory investigation to demonstrate conceptual understanding, I have common performance categories such as "purpose of investigation," "hypothesis," "procedures," "data transformations" (data tables and graphs), and "conclusions." In such authentic assessment tasks, I often state the problem and provide a partial list of available supplies and equipment that students may use. From this, students are to design and complete an experiment, and explain the results using scientific concepts and principles. The scoring guide template that I use for these authentic assessment tasks is illustrated in Table 1.

Table 1. Scoring Guide Template for Laboratory-Based Assessment Task

---

Assessment task heading
3 pts. Name, period, row and date all correct.
2 pts. Only three of four listed correct.
1 pt. Only two of four listed correct.
0 pts. Less than two of four listed correct.

Purpose of assessment task
(matches ALL aspects of the investigation)
3 pts. Purpose matches ALL aspects of the laboratory investigation.
2 pts. Purpose partially matches laboratory investigation.
1 pt. Does not match laboratory investigation.
0 pts. Not attempted.

Hypothesis for assessment task
(ACCURATE information that MATCHES the purpose)
3 pts. Hypothesis matches ALL information and the stated purpose.
2 pts. Partially matches information.
1 pt. Incorrect or incomplete.
0 pts. Not attempted.

Procedure for assessment task investigation
(ACCURATE information, COMPLETE, CLEAR STEPS)
10 pts. All steps were listed and necessary for completing the investigation.
8 pts. 95% - 80% of the steps were listed and necessary for completing the investigation.
6 pts. 79%-60% of the steps were listed and necessary for completing the investigation.
3 pts. 59% -30% of the steps were listed and necessary for completing the investigation, but more information is needed to complete the investigation.
0 pts. Less than 30% of the steps were listed and necessary for completing the investigation, OR NOT attempted

---

Table 1 Continues

Table 1. Scoring Guide Template for Laboratory-Based Assessment Task,
Continued

Procedure for using and naming/labeling equipment
4 pts. Accurately directs you to use the proper equipment for an investigation.
3 pts. Matches SOME proper equipment for an investigation.
0 pts. Not attempted.

Data reporting
(design of table or graph)
2 pts. Table or graph is complete, accurate, and correctly labeled.
1 pt. Table or graph is incomplete, inaccurate, and labeled incorrectly or
       incompletely.
0 pts. Not attempted.

Data transformation
(accuracy of table or graph)
2 pts. Table or graph accurate and completed with reasonable data.
1 pt. Table or graph completed but data is NOT reasonable or accurate.
0 pts. Not attempted.

Assessment task conclusion
(SPECIFIC to purpose, data, and procedure)
4 pts. Indicates ALL data ON WHICH the conclusion is based.
3 pts. Indicates SOME data ON WHICH the conclusion is based.
0 pts. No specific data mentioned OR not attempted.

Assessment task conclusion related to hypothesis
4 pts. Restates a correct form of hypothesis.
3 pts. Similar to hypothesis in form.
0 pts. Not attempted.

Assessment task conclusion accuracy
5 pts. ALL information is accurate and reasonable, and based on the
       investigation.
4 pts. Information is MORE than half accurate and reasonable, and based on the
       investigation.
3 pts. Information is LESS than half accurate and reasonable, and based on the
       investigation.
0 pts. Not accurate, reasonable, and based on the investigation OR not attempted.

Many problems available in the curriculum lend themselves to authentic assessment tasks. Some of these may stem from student questions. One of my favorites was asking students to explain how a fan helps to keep a person cool. Students believed that the air blowing from a fan was of a lower temperature. This student-generated question produced the assessment task as a apart of our study of relative humidity.

Based on my experiences, I offer the following for teachers who are developing authentic assessment tasks:

- Review your curricular and instructional goals. What are the major science concepts and science process and inquiry skills you want students to learn?
- Identify a potential assessment task. Review science textbooks and laboratory manuals for possible tasks, use student questions and ideas, look through magazines, journals, and newspaper articles for current issues and problems, and think of real world experiences.
- Determine whether the assessment task centers on the major science concepts, science process, and inquiry skills you want students to learn. Does the task lend itself to a meaningful context and real-world situations or problems? Do students use and apply science concepts to explain the task instead of simply recalling science concepts? Does the task integrate science concepts with science process and inquiry skills? Does the task require students to create a product?
- Write the assessment task to include a background context or scenario, a prompt to guide students' activity and response, and a list of materials and equipment needed.
- Design the scoring system based on the performance categories to be assessed. First, identify the specific science concepts and science process and inquiry skills to be assessed (i.e., the performance categories). Second, describe the criteria for each performance category to be used. Third, decide on the point range for each performance category.
- Field test, reflect on, and revise the assessment task: are there ways to improve the task? What did students think of the task? Did the task assess what you wanted? Was the scoring system effective and efficient?
- Keep several anonymous examples of student work to illustrate the nature and quality of the authentic assessment tasks for administrators, parents, and students.
- Do not limit yourself to assessment tasks that are completed in one class period. I often use tasks that require anywhere from two to five days of both in-class and out-of-class work.

One last suggestion about the design of authentic assessment tasks is to consider when the task is to be completed by individual students or by a group of students. I try to incorporate an equal number of individual and group assessment tasks. Although the authentic assessment tasks are designed to assess student performance, they also serve as learning tasks.

In implementing authentic assessment tasks in my classroom, I have found four conditions that must be met for the successful use of authentic assessments: (a) collaboration, (b) access to tools and resources, (c) discretion or opportunity for ownership, and (d) flexible use of time (Newmann, 1991). Collaboration refers to the opportunity for students to ask questions and receive feedback and assistance from peers, the teacher, and other resources. Access to tools and resources means making these available to students as they complete the task. Ownership involves students in conceiving the product or process, in executing the procedures, and in evaluating the product. This requires student control over the assessment task as students determine the procedures they will follow, ask questions, seek resources, and use language. Flexible use of time requires that students determine how to best use their time to complete the assessment task, instead of limiting work to a structured time period.

STUDENT USE OF SCORING GUIDES: SELF AND PEER ASSESSMENT

I have found that, when faced with authentic assessment tasks, students are able to achieve at higher levels if they are aware of the expectations and understand how they will be assessed. Therefore, I have found that scoring guides need to be made available to students during authentic assessment tasks. The more flexible the authentic assessment task, the greater the need to ensure that students understand the expectations. Providing the scoring guide with the assessment task assures that teacher and student share an understanding of these expectations and the degree of importance placed on each expectation.

As students complete the design for their assessment task, they use the scoring guide to evaluate their products and then to evaluate the products of others, a peer review process, before I assess the students' work. This provides a sense of responsibility and motivation for learning (AAAS, 1998). To earn a better score, students may alter their investigative designs or explanations following the peer assessment. This provides an opportunity to learn from the assessment task, an important aspect of the National Science Education Standards (NRC, 1996). My scoring guide has evolved and become more valuable to students as they became more familiar with using it to self- and peer-assess. This assessment process reflects the National Science Education Standards for developing self-directed learners. Students should be able to:

- Critique a sample of their own work using the teacher's standards and criteria for quality.
- Critique the work of other students in constructive ways (NRC, 1996, p. 88).

At first students may not be successful at using the scoring guides in these ways. It took several months of exposure to authentic assessment tasks and the accompanying scoring guide before my students became more comfortable with using the scoring guide on a regular basis. My first attempt at using the scoring guide to self-assess was not very successful; I discovered that students really did not critically self-assess or often failed to do so. I reinforced the idea that students could score their own work according to the scoring guide before submitting the assessment task. I model the use of a scoring guide to assess a previous student's product. This illustrates the standard for good work (AAAS, 1998) and facilitates students use of the scoring guides, but students need time and practice before they can effectively use the scoring guides to self- and peer-assess. I require students to score their own assessment tasks and then share the self-assessment with me. As students become more experienced, I spend less time assisting with the self-assessment process. As students become more familiar with the role of the scoring guide in classroom assessment, they begin to check the guide prior to handing in their responses. I have seen the quality of students' work greatly improve over time because of the self and peer-assessment process. This level of awareness has taken longer than I expected, but by mid-year students regularly compare their work to the scoring guides as a way of guiding their own activity.

As the year progressed, many students actually questioned the scoring guides I made available to them before the authentic assessment task. Students immediately questioned the science concepts and vocabulary used. I directed students back to class discussion, to any notes they took during lecture, to the textbook and other resources we used, and to our past laboratory investigations. The students were encouraged to check with their peers and to generate questions about the experimental design. These interactions generated much discussion and increased student understanding of the related concepts, vocabulary, and design of the investigation. According to Newmann (1991) such discussions encourage students to formulate and ask questions, to explain their ideas and procedures, and to refine their ideas and thinking. This differs from traditional classroom talk aimed at simply dispensing knowledge and evaluating students.

EXAMPLE OF AUTHENTIC ASSESSMENT TASKS AND TEACHER REFLECTIONS

In this section, I share two authentic assessment tasks (i.e., "A Mountain Emergency" and "Space Trip Trouble"), their scoring guides, and my self-assessment and reflection on each assessment task. The tasks illustrate the implementation of authentic assessment in science and the importance of self-assessment in the development of classroom assessments. These tasks reflect the assessment strategies proposed by AAAS (1998) in that they emphasize students' abilities to measure accurately, to use mathematics, to design experiments, and to solve problems. Students must also generate data and answers instead of simply recalling science facts.

"A Mountain Emergency"

"A Mountain Emergency" was developed as an authentic assessment task for an introductory science class (ninth to tenth grade). The purposes of this assessment task were to: evaluate students, assign a grade to student work, assess students' knowledge and understanding of distillation, evaporation, and condensation and provide feedback about instruction. The task develops problem-solving skills by allowing students to design and perform an experiment using equipment. To solve the problem, students must have a basic understanding of the concepts of distillation, evaporation, condensation, distilled water, and contaminants (in water). In completing the task, students utilize scientific reasoning skills, although these are not directly assessed by the task. The assessment task may be administered in either a small group or an individual setting, as a problem-solving format and then as a laboratory inquiry. This assessment task might also be given as a take-home practical in which students must return with a laboratory set-up and explanation, and with the equipment they would use. "A Mountain Emergency" assessment task follows:

What a time for this to happen! Here I am with my brother, returning from a nice two days at a friend's hunting cabin. The day started with a beautiful sunrise, great weather. A lovely wooded trail along a small, muddy stream back to the small village at the base of the mountains. We had only gone about halfway, maybe 20 miles, when trouble began. That angry she-bear (sow) chased us quite a distance until she decided that we really weren't a threat to her cubs. We dropped our packs when the chase began and that probably saved us as the bear stopped to explore and then destroy our packs and their contents. All I could save was this old teakettle, my pocketknife, a book of matches, and my plastic canteen. My brother had only the med-pack on his belt along with his car keys and some money. I thought we would hike on after we circumvented the bear and cubs when my brother had a medical emergency.

He needed an injection of his medication to stave off a heart attack. Unfortunately, all he had left was the medicine in pill form, a clean syringe with needle, a sterile plastic mixing bag, and a small measuring cup. The bear had broken his bottle of distilled water. What could I do to help? Only five milliliters of distilled water were needed. We had plenty of water from the stream, but no distillation apparatus to remove any microscopic contaminants, and certainly no tubing of any sort. I tried to make my brother as comfortable as I could under the shade of the tree. It was beautiful, without a cloud in the sky, but there was no help for miles around. I searched my pockets for anything that would help. In my brother's jacket, I found a new teabag. As I looked at this, at my teakettle, and the contents of my brother's med kit, I tried to think of an answer. There would not be enough time to get help. All I needed was five milliliters of distilled water to make the solution to save my brother.

I thought I would make him a cup of hot tea but I didn't have a cup. Of course, the canteen could serve as a cup. In a situation like this, one has to improvise. Suddenly, I had the answer: I could distill a little water with what I had available.

Purpose: design a means of producing five milliliters of distilled water for the medical emergency given the materials available. Draw your set-up, label its parts, and explain why it would cause evaporation and condensation, producing distilled water. Think back to the experience you gained by distilling the water in class and to any other experiences you might have had when water vapor condensed into a liquid. Your work will be judged by how well it could distill water and remove contaminants using only the materials at hand. Have your paper initialed by the teacher and then go to the lab tables and try it. If you make any changes, draw, label, and explain your new set-up and the scientific processes involved. For guidance, check the scoring guide!

The scoring guide for the assessment task reflects students' ability to distill water, to design a distillation set-up using only the materials provided, and to use scientific concepts to explain the distillation process. The scoring guide is displayed in Table 2.

Reflection on "A Mountain Emergency"

The purpose of this authentic assessment task was to assess students' understanding of distillation, as well as their ability to design, complete, and report on a unique problem. The students utilized skills gained from instructional activities that focused on determining the boiling point of a liquid and on fractional distillation. The students worked in groups to complete the task. I placed the laboratory supplies out in the room with labels indicating what role they played in the story. All supplies were available even though not all were needed. The assessment task was successful for the average and the advanced student. Both classes were able to meet all of the criteria and to complete the investigation by solving the problem.

Table 2. Scoring Guide for "A Mountain Emergency"

Performance Category	Points
1. Student has work okayed by teacher before starting lab work	5 pts.
2. Following lab set-up, the student recovers:	
5 mL of uncontaminated distilled water	10 pts.
5 mL distilled water but contaminated	7 pts.
Less than 5 mL distilled or contaminated water	5 pts.
3. Student only used materials available in story	5 pts.
Used one extra thing needed but not part of story	3 pts.
Used several extra items needed but not part of story	2 pts.
4. Drawing labels are complete including all equipment, location of related terms (condensation, evaporation, distilled water, and contaminants), and materials involved with this lab to show distillation.	5 pts.
Drawing is incomplete or partially labeled	3 pts.
Drawing is incomplete or not labeled	2 pts.
Drawing is attempted, but too much is missing to allow interpretation	1 pt.
5. The student relates lab set-up to the following terms to explain the distillation process that occurs: condensation, evaporation, distilled water, contaminants	5 pts.
Only 2 or 3 of the science concepts are correctly used or appropriately related to drawing.	3 pts.
Only 2 or 3 of the science concepts are correctly used, but are not appropriately related to drawing.	2 pts.
Fewer than 2 science concepts are correctly used or none are appropriately related to the drawing.	1 pt.
Total possible points	30

To improve this assessment task, I changed the manner in which I administered it to students. The first class to try this assessment seemed to find only one right answer. I felt that it was too easy for the students because they could see what other students had done instead of trying an original approach. The students in essence were copying each other. Subsequent classes produced varied approaches and acceptable answers because I required them to shield their materials and experimental set-up from the other student groups using wooden pegboards. By stressing originality and giving students an opportunity to hide their ideas from others, they spent more time trying to solve the problem their own way instead of looking around for ideas. I felt this was more useful. I evaluated the assessment task using the evaluation matrix (see Appendix A) developed in the Eisenhower supported assessment project (Shepardson, 1996).

"Space Trip Trouble"

I use "Space Trip Trouble" to evaluate students and assign a grade, and to provide feedback on my instruction. The task assesses students' knowledge about boiling points, fractional distillation, evaporation, and condensation. Students are engaged in solving a problem by designing and performing an investigation using equipment found in the laboratory. To be successful at this task, students must understand the concepts of boiling point, fractional distillation, evaporation, and condensation. This task engages students in scientific reasoning, but does not specifically assess this reasoning.

The assessment task may be administered in either small-group or individual settings, as a problem-solving format, and then as a laboratory inquiry. It is used in conjunction with information available to students describing fractional distillation. Students are provided with samples of the pure materials that are in the solution so that they can test and identify the materials and use that information to successfully design their investigations. The scoring guide is displayed in Table 3, and I also administer a more traditional quiz (Table 4). The assessment task:

> On the first space flight to Mars, you have encountered your first flight crisis. During routine maintenance, the fuel line was accidentally connected to your water supply line. The water mixed with the denatured alcohol used as a source of fuel for your emergency steering rockets. The rockets can operate temporarily, but your water supply is completely contaminated. Solar flare activity has interrupted radio contact with Earth. You need to develop a means of separating the fuel from the water and returning both to their proper containers. Fortunately, several empty water and fuel canisters are available if you can develop the correct experimental plan to save yourself and the three billion dollar mission to Mars. What are you going to do?

> Available to each research team: small samples of pure water and fuel to test and record any appropriate characteristic properties we have none to spare, a sample of the fuel-water solution, and access to and assistance in the laboratory. TIME IS RUNNING OUT! You must develop your experimental procedure and PROVE you have pure samples within three class periods before all emergency supplies are exhausted and everyone aboard your spacecraft perishes!

Required:

- Research: make a list of sources of information available to you that includes a numbered list of important facts, concepts, and relationships, that may help you solve your problem. (Develop each of these according to past requirements in class.)
- Purpose
- Research: Information from book, short labs with data
- Hypothesis
- Procedure (use story board type)
- Data
- Teacher's initials to start experiment: _____
- Conclusion report: indicate how successful you were, any reasons why your results were not pure, and how you tested your products to ensure purity.

Reflection on "Space Trip Trouble"

The purpose of this authentic assessment task was to introduce the chapter on separations in the introductory physical science text. Students were to demonstrate previously learned skills by determining the boiling point and density of a liquid. These skills were to be shown in the research section of this assessment and students were to use information gained to help design and complete the investigation. This assessment activity was successful for the advanced students who were able to meet all of the criteria and complete the investigation. The average students were confused; this was the first time they had experienced a research section, including small experiments. As a result, many of the students wrote their conclusions after completing the research section and writing their hypotheses. Although they completed the work in class, they did not include it in their laboratory reports.

To improve this assessment task, I would divide it into separate parts with daily deadlines. The students had three class periods to research, design and complete their experiments. Some students could identify what they were doing and what they had discovered about the unknown liquid (their "rocket fuel"), but they seemed to lose track of their written work. Perhaps this was because they had a checklist but not the scoring guide. I had encountered some difficulties in designing a scoring guide that would not give away too much information, so I tried a checklist. In the future I will use separate lists for each part of the task and include point values for each. Nonetheless, the students seemed to enjoy this assessment task and worked well together. Some had a hard time getting started, and I attributed this to the length of time available. I wondered if students felt they had time to waste at the beginning of the task because they had three periods to finish. My self-assessment of the task, based on the assessment evaluation matrix, is shown in Appendix B.

Table 3. Scoring Guide for "Space Trip Trouble"

Performance Category	Completely correct	Partially correct	Incomplete or wrong
Team members listed	2	1	0
Research 1. Lab research on fuel boiling point: includes correct & complete purpose, storyboard procedure, data, conclusion statement (+2 each) 2. Lab research on fuel density: includes correct & complete purpose, storyboard procedures, data conclusion statement (+2 each)	8 8	6 to 2 6 to 2	0 0
Hypothesis (based on research) 1. States correct full name 2. Uses of fractional distillation to separate	2 2	1 1	0 0
Procedure 1. Had teacher OK before starting 2. Uses drawings appropriately 3. Includes entire procedure (+1 each)	2 2 6	1 1 5 to 1	0 0 0
Data 1. Had teacher OK before starting 2. Mass of G.C., G.C. + distilled fuel, distilled fuel, G.C. + distilled water (+1 each) 3. Density of distilled fuel, distilled mixture, distilled water (+2 each) 4. Time & temperature table 5. Boiling points of distilled fuel, distilled water (+2 each)	2 4 6 4 4	1 3 to 1 4 to 2 3 to 1 2	0 0 0 0 0
Conclusion 1. States name of fuel & data proving it (+4 each) 2. States means of separation based on boiling point (+4 each)	8 8	4 4	0 0

Table 4. Example Quiz on Separating Fuel and Water

---

1. Which of the three alcohols (methyl alcohol, ethyl alcohol, or isopropyl alcohol) did you identify as the "fuel" in the problem?
2. What was the name of the process used to separate these materials?
3. Describe and explain in your own words what would happen to the temperature at which a test tube of water would boil if you lowered the test tube closer to the flame on your Bunsen Burner.
4. Identify what you thought was to be collected in the first test tube, the second test tube, and the third test tube sitting in the cold water.
5. EXPLAIN in your own words what happens to the heat energy from the Bunsen Burner if a liquid is boiling. Include what happens to the temperature of the liquid before it boils! Tell what happens to the temperature of the liquid while it is boiling! (You may use illustrations to help explain your ideas, but be certain you label the parts.)

CONCLUSION

Authentic assessments provide opportunities for students to solve problems. The manner and degree to which they solve these problems and explain their solutions provide tremendous insight into their level of understanding. Authentic assessments are not only a means of assessing students, but also serve to reinforce knowledge, understanding, and student exploration. Authentic assessment motivates students and cultivates their higher-order thinking and problem-solving capacities beyond a preconceived level of minimums (Newmann, 1991). I wanted to raise student awareness about my expectations for class work. Although I had used scoring guides to assess student work after it had been collected, it was only recently that I realized students would benefit from exposure to the scoring guide as a part of the authentic assessment task. I believe that some of the students have become more critical of their work and have improved the level of their class work in science. Eventually, I hope that this will elevate students' expectations and responsibility, thus improving their science learning.

REFERENCES

American Association for the Advancement of Science (1990). *Science for all Americans*. New York: Oxford University Press.

American Association for the Advancement of Science (1998). *Blueprints for reform: Science, mathematics, and technology education*. New York: Oxford University Press.

Baron, J.B. (1991). Performance assessment: Blurring the edges of assessment, curriculum, and instruction. In G. Kulm and S.M. Malcom (Eds.), *Science assessment in the service of reform* (pp. 247-266). Washington, DC: American Association for the Advancement of Science.

Brooks, J.G. & Brooks, M.G. (1993). *In search of understanding: The case for constructivist classrooms*. Alexandria, VA: Association for Supervision and Curriculum Development.

Newmann, F.M. (1991). Linking restructuring to authentic student achievement. *Phi Delta Kappan*, 73(6), 458-463.

National Research Council (1996). *National Science Education Standards*. Washington, DC: National
    Academy Press.
Shepardson, D.P. (1996). Teacher enhancement through alternative assessment task development. A
    proposal funded by the State of Indiana Commission for Higher Education.

APPENDIX A: ASSESSMENT EVALUATION MATRIX FOR "A MOUNTAIN EMERGENCY"

Rating Scheme
E = exceptional, specifically or clearly addressed.
NS = not shown.
S = shown, but not specifically or clearly addressed.

Purpose   E   S   NS

Clear and useful to students because students translated the purpose
from the student paper into a simpler form that is approved by the
teacher.

	E	S	NS
Is the purpose of the assessment clearly articulated?	x		
Does the assessment provide useful information to the shareholders?	x		
Is the assessment a part of a repertoire of a wide range of performances	x		

Validity

The assessment and purpose are directly related.   The student
communicates knowledge within the procedure and the conclusion.

	E	S	NS
Is there a match between the purpose of the assessment and the process used?	x		
Does the assessment measure valued outcomes?	x		
Key themes or concepts	x		
Cognitive skills		x	
Does the assessment provide an opportunity for the student to communicate what he/she knows?	x		
Does the assessment provide information for logical and sound inferences?	x		
Is the written plan for the assessment purposefully designed?	x		

Learning Tool

The assessment is ongoing and appropriate.   Students have the
opportunity to change papers and work (peer evaluation) and self-
evaluate based on the scoring guide developed from student input.

	E	S	NS
Is the assessment an ongoing process?	x		
Is the assessment developmentally appropriate for the students with whom it will be used?	x		
Is there an opportunity for students to engage in			

Self assessment	X		
Peer assessment	X		

Authenticity

Very good. Relative to real world problem in a context that requires students to apply their science understandings. I was interested in having students decide if and how they could save their brother by completing this investigation.

Does the assessment reflect science as scientists do science?		X	
Does the assessment reflect the real world?	X		
Does the assessment engage students' interests?	X		
Does the assessment require the student to demonstrate knowledge to make personal and societal decisions?	X		
Is the assessment inquiry-based?	X		

Equity

Not measured, but students are provided the opportunity and are expected to refer to past work for help.

Is the assessment equitable?	X		
Is there evidence that the students have had the opportunity to learn the content and skills covered by the assessment?		X	

Technical Aspects

Prompt clearly written and used by students without difficulty.

Does the assessment task have a clearly written prompt?	X		
Are the criteria for scoring the assessment provided with the task?	X		
Are the scoring criteria appropriate for the task?	X		
Do the scoring criteria define a wide range of performances?	X		

APPENDIX B: ASSESSMENT EVALUATION MATRIX FOR "SPACE TRIP TROUBLE"

Rating Scheme

E = exceptional, specifically or clearly addressed.
S = shown, but not specifically or clearly addressed.
NS = not shown.

Purpose   E   S   NS

The purpose is developed by the students based on the prompt, but this must be approved by the teacher before students start the procedure. This gives students a chance to determine the purpose, verify that their ideas match the task and understand how the task will be scored.

Is the purpose of the assessment clearly articulated?	X		
Does the assessment provide useful information to the	X		

shareholders?			
Is the assessment a part of a repertoire of a wide range of performances	x		

Validity

I believe there is a very close match between the purpose and the assessment task. The students demonstrated their understanding by successfully completing the investigation. Some of the valued outcomes are not directly listed because these are skills that must be demonstrated for successful completion of the end product.

Is there a match between the purpose of the assessment and the process used?	x		
Does the assessment measure valued outcomes?		x	
Key themes or concepts		x	
Cognitive skills			x
Does the assessment provide an opportunity for the student to communicate what he/she knows?	x		
Does the assessment provide information for logical and sound inferences?	x		
Is the written plan for the assessment purposefully designed?	x		

Learning Tool

Students realize that this is dependent upon past class work and is another opportunity to show they have learned. By having students take time to check, correct, improve their papers based on an expected outcome list they have the opportunity to self-assess, but this part needs improvement. I have some difficulty with peer assessment in this class, so I have dropped it for now. We will continue to develop students' abilities to peer-evaluate.

Is the assessment an ongoing process?	x		
Is the assessment developmentally appropriate for the students with whom it will be used?	x		
Is there an opportunity for students to engage in			
Self assessment	x		
Peer assessment			x

Authenticity

Students realize that they will probably not go on a trip to Mars, but through class discussion and questions demonstrate that they realize that this is a possible problem for a scientist or astronaut. Students are interested, and it did not seem appropriate for personal or societal decisions to be included.

Does the assessment reflect science as scientists do science?	x		

Does the assessment reflect the real world?		X	
Does the assessment engage students' interests?	X		
Does the assessment require the student to demonstrate knowledge to make personal and societal decisions?			X
Is the assessment inquiry-based?	X		

Equity

Not measured, but students are expected to return to past investigations for help in their research section.

Is the assessment equitable?	X		
Is there evidence that the students have had the opportunity to learn the content and skills covered by the assessment?	X		

Technical Aspects

Prompt is clear to students, but the scoring guide is not provided to them; instead a checklist of outcomes is provided. Students need more direction here.

Does the assessment task have a clearly written prompt?	X		
Are the criteria for scoring the assessment provided with the task?	X		
Are the scoring criteria appropriate for the task?	X		
Do the scoring criteria define a wide range of performances?	X		

DANIEL P. SHEPARDSON

INTRODUCTION TO SECTION III: COMMON THEMES

The last chapter in this book summarizes the common themes about assessment in science. The first section of the chapter elucidates the successful strategies for professional development and identifies the common issues that professional development programs must attend to in order to successfully challenge and change classroom assessment practice. The second section of the chapter synthesizes the common elements of the teachers' chapters, articulating aspects of successful classroom practice. The chapter concludes by calling for long-term support of professional development programs in order to change teachers' assessment in science.

Daniel P. Shepardson (ed), Assessment in Science, 249.
© 2001 *Kluwer Academic Publishers. Printed in the Netherlands.*

DANIEL P. SHEPARDSON

15. WHAT WE HAVE LEARNED: A SUMMARY OF CHAPTER THEMES

In this closing chapter, I summarize the key ideas and issues surrounding professional development in science assessment and classroom assessment practice. It will be difficult, however, to do justice to all of the ideas, examples, tools, and issues presented by the authors in the previous chapters. The best way may be to highlight the essential ingredients, the common themes across the chapters. I present these themes in two sections: first, professional development in science assessment and, second, teachers' classroom assessment practice. The chapter concludes with a thought about long-term support for changing classroom assessment practice.

PROFESSIONAL DEVELOPMENT IN SCIENCE ASSESSMENT

Professional development programs that change science teachers' assessment practices require tremendous energy, time, resources, and a long-term commitment to change. Collaboration is also essential to professional development programs in science assessment, collaboration among staff developers and classroom teachers and collaboration among classroom teachers. The essence of change lies in challenging teachers' understanding and knowledge about their assessment practice. Consistent with this notion is the process of engaging teachers in critically reflecting on their classroom assessment practice. This must go beyond simply documenting classroom practice to include the critical analysis of this practice in the light of current assessment standards or alternative ways of knowing. In this way, reflection supports the modification of practice based on new understandings and knowledge about assessment; this results in the production of new tools or ways of assessing in the context of each teacher's classroom and school environment. Many of the teacher chapters in this book describe how critical reflection on classroom assessments resulted in changes in both instruction and assessment practices.

Teachers begin by constructing a current picture or profile of their assessment practice and then move toward a consensual understanding of what assessment might look like in a science classroom. The key element is the development of a clear, coherent, and common meaning about what change in practice is for, what it involves, and how it will happen (Fullan, 1991). As a part of this process, teachers experience an assessment task that they not only critically analyze from their own

Daniel P. Shepardson (ed.), Assessment in Science, 251—255.
© 2001 *Kluwer Academic Publishers. Printed in the Netherlands.*

perspectives, but also from those of the reform documents in science education (e.g., *National Science Education Standards* and *Benchmarks for Science Literacy*).

Next, professional development programs must engage teachers in using the reform documents to construct new understandings, new tools for reflection and analysis, and new products of practice. When used as tools for reflecting on current practice and guiding the development of new practice, the science education reform documents contribute to changing classroom practice. When used as administrative policy pieces, however, they tend to have little impact on classroom practice. These new understandings, new tools, and new products must then be put to the classroom test and used in the reflection process. A critical component of reflection, analysis, and development is the interaction and sharing among peers.

Although many issues must be considered in designing and implementing professional development programs in science assessment, key constraints include time, instruction, stakeholders, students, learning, support, classroom context, and collaboration. Without considering these issues, professional development programs are likely to have limited success in changing classroom practice. An often overlooked aspect of the development of classroom-based assessment is the research on children's ideas. Classroom assessments based on the ideas and understandings of children not only better inform teachers about children's science learning, but also allow for progression in children's understandings to be monitored, better informing instruction. Children's developmental levels and experiences must be taken into consideration in the design and implementation of assessment tasks.

Assessment practice is shaped not only by teacher knowledge and beliefs about assessment, but also by knowledge about pedagogy, science content, and about learners. Enhancing teachers' assessment knowledge alone will not necessarily lead to change in classroom assessment practice. Professional development programs must address science assessment in the context of the teachers' classroom; a context shaped by the science content (curriculum), by pedagogy, and by the learner, or pedagogical-content-assessment knowledge. Engaging teachers in analyzing student work and classroom assessment data provides the foundation and context for changing assessment practice and the evidence for improving instruction. The interpretation of student assessment data links assessment, pedagogy, and learning. In essence, professional development programs engage teachers in constructing classroom assessment tasks, collecting student work examples based on the developed assessment tasks—assessment data as evidence, interpreting the student assessment data to inform classroom practice, and then changing classroom practice based on student assessment data.

TEACHER'S CLASSROOM ASSESSMENT PRACTICE

As teachers we often think of classroom activities as either being pedagogical in nature or as stand-a-lone assessment tasks. Pedagogical tasks are viewed for the purpose of mediating students' science learning and assessment tasks are seen as a means for measuring or determining what students have learned. One theme that has connected these chapters is the notion of pedagogical-assessment tasks. The idea that assessment need not stand alone, separate from instruction, as an isolated event at the end of the instructional sequence. Pedagogical-assessment tasks foster a way of thinking about classroom activities as serving the purpose of teaching, learning, and assessing. It promotes a way of thinking about practice as it raises questions about practice: "How might this activity be used to mediate students learning and serve as a means for assessing students?" "How might this activity be changed to both teach students and assess their performance?" Assessment in this way is not an afterthought to instruction, but is built into the design of instruction (NRC, 1996).

Pedagogical-assessment tasks not only align assessment with instruction, but also embed assessment in instruction, providing students with an opportunity to learn science (NRC, 1996) as they are assessed. Children are assessed as they engage in learning, not at the conclusion of the learning process. Good assessments are also good learning experiences (NRC, 1996). In pedagogical-assessment tasks, teachers mediate students' activity through questions and suggestions, making available resources for completing the task. Through this process, the activity provides students with an opportunity to learn. Student processes and products become the indicators of performance in science. Pedagogical-assessment activities also better inform practice by providing teachers with immediate feedback about student progress and instruction, providing the teacher with the opportunity to make different pedagogical moves.

Consistent with the pedagogical-assessment theme is the view that assessment serves the purpose of diagnosing students' science understandings and abilities. The variety of assessment tools presented throughout this book provide ways of documenting children's prior understandings and abilities. Many of the examples emphasize students' exploration and manipulation of phenomena and equipment, resulting in the creation of a product that communicates understanding. These examples also emphasize the importance of assessing students' abilities to develop investigative plans, equally stressing planning, experimenting, and knowing science concepts beyond simply knowing science facts. Although Emery's chapter specifically discusses authentic assessment, many of the assessments shared in this book also reflect classroom-based authenticity.

Changing classroom assessment practice often results in changing the curriculum and instruction. The teacher chapters in this book clearly illustrate how these teachers changed their classroom curriculum and instructional activities as a result of either developing new assessments or interpreting student assessment data. Through improved assessment practice, these teachers gained new and better insight into their students' science understandings and abilities. Further, because these teachers used the classroom assessment data to go beyond evaluating and grading students, they were able to make more informed decisions about their curriculum and instructional activities. In turn, this improved student learning. These teachers' assessment practice changed from grading and evaluating students to an emphasis on learn about what students know and can do in science (NRC, 1996).

Children's drawing, writing, and talking are threads that weave their way through the chapters in Section Two. Assessment tools based on these graphic forms capture and chronicle children's developing science understandings and abilities. They profile children's progression in both science learning and literacy development, and embed assessment in instruction. Children's talk also reveals patterns in understanding that inform pedagogy. At the same time, however, teachers must understand the importance of their own talk and its influence on children's action and understanding. The questions teachers ask mediate children's science learning.

Suggestions for using children's writing and drawing as teaching, learning, and assessment tools are presented throughout the book. These assessment examples go beyond simply assessing students' content understandings and recording of observations to look at conceptual understanding, science processes, attitude, and the use of language. Children's writing and drawing also convey different world views. These world views may be reflected in the organization of children's writing and drawing by engaging in scientific discourse or carrying out scientific procedures. Different children use different world views to make meaning of school science activities. Understanding these is essential to viewing children's graphic products as assessment tools and tailoring instruction to meet the needs of children.

What should be assessed in science classrooms? Although the answer to this question is likely to stir debate, the themes that have emerged across these chapters include the domains of:

- Discipline-based knowledge
- Science processes and inquiry skills
- Reasoning processes
- Metacognitve processes

Performance standards for these domains may be located in the science education reform documents and literature. The NRC (1996) *National Science Education Standards* and AAAS (1993) *Benchmarks* can guide assessment design and provide performance criteria. The chapters authored by teachers articulate the ways in which the reform documents have contributed to the development of assessment tasks whether through the identification of science content to be assessed or the construction of a scoring guide. What also emerges from these chapters is the notion that what is assessed depends upon teachers' views of science and what they value in science. As Britsch sums it, "our assessments reflect what *counts* in each of our classrooms" (2001, p. 111).

Revision, revision, revision. This echoes throughout the teacher chapters. The teachers' first attempts at developing and implementing assessment tasks in their classrooms most often resulted in the need to revise the tasks. No assessment task was deemed perfect in its first use. In response, the teachers found out what students thought about the assessment tasks, and used these perspectives to guide the revision. Student practice and familiarity with the assessment task format laid the groundwork for success. In other words, students needed opportunities to learn to "take" the assessments. Changing assessment practice also requires changing student views of assessment. Of all of the suggestions for changing assessment practice and developing assessment tasks presented by these teachers, the most important is to change gradually. Too many changes at once may be overwhelming and frustrating, leading to no change at all.

Perhaps the most important change shared by these teachers is their increased knowledge about students' understandings and abilities as compared to the use of more traditional methods of assessment. For all of these teachers, assessment has become an integral component of classroom instruction and an opportunity for students to learn science. These classroom assessments inform teachers about student learning and about their teaching. Changing assessment practice was not easy for these teachers, requiring considerable time and energy, but the benefits of knowing more about their students and teaching far outweighed these costs.

CLOSING THOUGHT

Many professional development strategies and tools have been presented here, and a wide range of classroom-based assessment examples have been shared. The staff developer and the classroom teacher must modify these strategies and tools to fit each specific situation. Simply plugging in an assessment activity for teachers or an assessment task in the classroom may not work. The design and use of science assessment tasks requires careful thought about the consequences, generalizability (Linn, Baker, & Dunbar, 1991), reliability, and validity of the assessment task or activity. Issues about designing multiple trait and reasoning-based assessment tasks must also be addressed in professional development programs. Classroom assessment must become integral to the reform of science education. In turn, school districts and state and federal agencies must be willing to support teachers over the long term instead of funding short-term fixes. Few teachers will change their assessment practice beyond the support of the professional development program. Long-term systemic change in science assessment is only possible through long-term support of this change.

REFERENCES

American Association for the Advancement of Science (1993). *Benchmarks for science literacy*. New York: Oxford University Press.

Britsch, S.J. (2001). Assessment for emergent science literacy in classrooms for young children. In D.P. Shepardson (Ed.), *Assessment in science: A guide to professional development and classroom practice* (p. 111). Dordrecht, The Netherlands: Kluwer Academic Publishers.

Fullan, M.G. (1991). *The new meaning of educational change*. New York: Teachers College Press.

Linn, R.L., Baker, E.L., & Dunbar, S.B. (1991). Complex, performance-based assessment: Expectations and validation criteria. *Educational Researcher*, 20(8), 15-23.

National Research Council. (1996). *National science education standards*. Washington, DC: National Academy Press.

NOTES ON CONTRIBUTORS

Susan J. Britsch received her Ph.D. from the University of California, Berkeley and is Associate Professor of Literacy and Language Education at Purdue University. Professor Britsch's most recent research examines the uses of electronic mail for literacy development in elementary school classrooms, and integration of journaling with elementary school science curricula.

Marilynn Edwards is a seventh-grade teacher at Taft Middle School in Crown Point, Indiana. Marilynn teaches five classes of life science each day with approximately 29 students per class. She is professionally active, participating in and conducting workshops for teachers and is involved in the state teachers association. Marilynn is also active in and presents at state and national science teacher conferences.

David Emery currently teaches introductory physical science courses at Elkhart High School after teaching eighth grade science at West Side Middle School, Elkhart Indiana. David has been extensively involved in professional development programs, and conducts workshops for teachers on computer technology and science teaching within the school corporation.

Vicky Flick is a high school chemistry teacher at Brownsburg High School, Bronwsburg, Indiana. She teaches six chemistry classes each day with the maximum of 24 students in each class. Vicky is in her 15th year of teaching. Vicky is professionally active, conducting workshops for teachers and presenting at state and national science teacher conferences.

Edith S. Gummer is a graduate of the Science Education program, Department of Curriculum and Instruction, Purdue University. She is currently an Assistant Professor of Science Education at Oregon State University. She was a former science teacher and department chair and consultant to the *New York State Science Assessment Project*. She has worked as the assistant to the Assessment Working Group Chair on the National Research Council, *National Science Education Standards* project and served as the Project Coordinator of the *Teacher Enhancement trough Alternative Assessment Task Development* project funded by the State of Indiana Commission for Higher Education.

Vicky Jackson is a seventh-grade science teacher at Lebanon Middle School, Lebanon, Indiana. She teaches six science classes each day, ranging between 20 and

30 students per class. The middle school emphasizes a teaming approach where grade level teams collaborate on planning curricular and instructional activities and themes.

Ted Leuenberger is a seventh grade science teacher at Benton Central Jr. - Sr. High School, Oxford, Indiana. He teaches within an academic team. He has 25 years of teaching experience, and is actively involved in professional development activities.

Brenda Main is an elementary teacher at Northwood Elementary School, Franklin, Indiana. In her 28 years as an elementary teacher she has taught grades 2-5 and special education. Eleven of those years were spent teaching 4-5 science classes at Northwood Elementary School. Brenda has also taught the science methods class for Elementary Education Majors at Franklin College.

Daniel P. Shepardson is an Associate Professor of Science Education, Purdue University. He received his Ph.D. in Science Education from the University of Iowa. Professor Shepardson has served as PI or Co-PI on numerous federal and state funded teacher enhancement projects dealing with assessment, laboratory and inquiry-based teaching, and environmental science. He has published extensively and presented at state, national, and international conferences on assessment, professional development, inquiry-based teaching, and children's science learning. He teaches undergraduate and graduate courses in science education and environmental education.

SUBJECT INDEX